THE OAKWOOD PRESS

Private Owner Wagons

by
Bill Hudson

THE OAKWOOD PRESS

British Library Cataloguing in Publication Data
A Record for this book is available from the British Library
ISBN 0 85361 492 X

Typeset by Oakwood Graphics.
Repro by Ford Graphics, Ringwood, Hants.
Printed by Alpha Print (Oxford) Ltd, Witney, Oxon.

Plate 1: This Plate has been chosen as a classic example of the 1907 RCH 10 ton mineral wagon. It measures 16 ft x 7 ft 6 in. x 4 ft 0 in. (external) and has 7 x 7 in. planks. (To be meticulous the bottom six are 6⅞ in. and the top one is 6¾ in.) The design is characterised by brakes on one side only, grease axleboxes (in this case Attock's 46), crown plates (the semi-circular pieces on the solebar), a metal plate on the door (to protect the woodwork when the latter is dropped) and plain buffer guides. It is fitted with side, end and bottom doors and the commonly used pivot bar end door fastening. It also has commode handles often fitted to Welsh wagons. The wagon was from a batch of 25 ordered in January 1908, numbered 293-317, and was painted red, with white letters, black shading (Precision Paints GWR China Red was the correct shade). Although the running gear is painted red, this was only for the benefit of the photographer, and would have been given a coat of black before the vehicle entered service.

Published by
The Oakwood Press
P.O. Box 122, Headington, Oxford, OX3 8LU.

Contents

Acknowledgements

Over the years many individuals have contributed to my knowledge of private owner wagons and to these people I again express my gratitude. In preparing this narrative I would particularly like to thank Mike Grocock, Keith Turner (Birmingham Central Library), and the staff of British Coal Archives, Somerset County Record Office, and Derbyshire Library Service. I am most grateful to Gordon Wiseman for permission to reproduce Sir Felix Pole's letter on page 51, courtesy *Railway Gazette International*. Finally, as usual, my thanks to Chris Crofts, who not only corrected my mistakes, but offered much constructive criticism and advice.

All plates are from author's collection except:

Plate 3	E. Meadowcroft
Plate 5, 11, 12, 19, 21, 23, 24, 25, 28, 30, 34, 37, 38, 41, 44, 61	HMRS
Plate 13	The Keeper, National Railway Museum
Plate 14	L. Perry
Plate 27	M.E. Morton-Lloyd
Plate 33	A. G. Johnson
Plate 38	G. Harvey
Plate 57	F.W. Shuttleworth

LONGITUDINAL SECTION

SIDE ELEVATION

SIDE & END SHEETING TO BE DEAL OR LARCH

TO BE LET INTO SIDE OF WAGON

CAPACITY = 534 CUBIC FEET

8'-6¼" FROM RAIL TO TOP OF CAPPING

4'-4½" DEPTH INSIDE

⅝" BOLTS

16'-6" OUTSIDE
16'-1¼" INSIDE
4'-6"

6'-0"

3'-5¼"

3'-9"

9'-0" WHEEL-BASE

2'-11¾"

DOOR OUTSIDE
DOORWAY OUTSIDE
DOOR INSIDE
DOORWAY INSIDE

4'-0½"
4'-1½"
3'-11½"
4'-0"

2¼" × ¼" PLATE

2⅜"
3⅜"

5" × 4" × ½" T

BEARING SPRING
5 PLATES 4" × ½"

WHEELS 3'-1½" DIA. ON TREAD WITH 2¼" TIRES

9" × 4½"

2⅝"

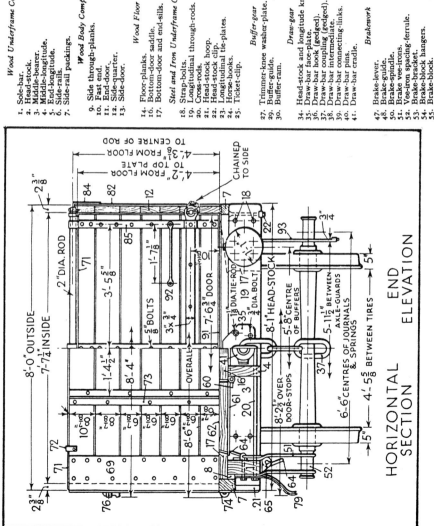

Wood Underframe Components
1. Sole-bar.
2. Head-stock.
3. Middle-bearer.
4. Middle-longitude.
5. End-longitude.
6. Side-rails.
7. Side-rail packings.

Wood Body Components
9. Side through-planks.
10. Fast end.
11. End-door.
12. Side-quarter.
13. Side-door.

Wood Floor
14. Floor-planks.
16. Bottom-door saddle.
17. Bottom-door and end-sills.

Steel and Iron Underframe Components
18. Strap-bolts.
19. Longitudinal through-rods.
20. Cross-rods.
21. Head-stock hoop.
22. Head-stock clip.
23. Longitudinal tie-plates.
24. Horse-hooks.
25. Ticket-clip.

Buffer-gear
27. Trimmer-knee washer-plate.
29. Buffer-guide.
30. Buffer-ram.

Draw-gear
34. Head-stock and longitude knee.
35. Draw-bar face-plate.
36. Draw-bar hook (gedged).
37. Draw-bar coupling (gedged).
38. Draw-bar intermediate.
39. Draw-bar connecting-links.
40. Draw-bar pins.
41. Draw-bar cradle.

Brakework
47. Brake-lever.
48. Brake-guide.
50. Brake-spindle.
51. Brake vee-irons.
52. Vee-iron spacing-ferrule.
53. Brake-bracket.
54. Brake-block hangers.
55. Brake-block.

56. Brake push-rods.
58. Brake-carriers.
59. Brake centre safety-loop.

Bottom-door
60. Hinge saddle-band.
61. Hinge.
62. Hinge washer-plate.
64. Monkey-tail.
65. Monkey-tail bracket.
66. Monkey-tail peg and chain.

Steel Body Components
67. Side-knees.
68. Side-knees washer-plates.
69. Corner-plates.
70. Diagonal side-braces.
71. Capping.
72. Capping-clip.
73. End stanchion.

Side-door.
74. Hinge.
75. Hinge washer-plate.
76. Hinge-fastener.
77. Hinge-fastener washer-plate.
78. Door striking-plate.
79. Door stop-spring.

End-door
81. End-door knees.
83. End-door roller-bar.
84. End-door roller-bar stop.
85. End-door hinges.
86. End-door hinges spacing-ferrule.
87. End-door hinges washer-plate.
89. End-door fastening-bar bolts.
90. End-door pegs and chains.
91. End-door sill-plate.
92. End-door commode-handle.

Carriage
93. Axle-guard.
94. Axle-guard bridle.
95. Axle-box top section.
96. Axle-box bottom section.
100. Wheels.
101. Side bearing-springs.
102. Side bearing-springs shoes.
103. Side bearing-springs stop.
104. Railway register-plate.

HORIZONTAL SECTION

END ELEVATION

Plate 2: By comparison with *Plate 1* this plate shows a perfect example of the 1923 RCH 8-plank mineral wagon. It is 16 ft 6 in. x 8 ft 0 in. x 4 ft 7 in., with 8 planks at 6⅞ in. width. It has brakes on both sides, split oil axleboxes, no crown plates, a short strap on the door to connect with the door bang stop on the solebar below, and ribbed buffer guides. Those at the door end have an upward extension to keep the end sill in place. By this time the end door fastening had become the pin and chain type. Allerton Main Colliery was situated at Kippax, on the ex-NER Garforth-Castleford line and should not be confused with the nearby Allerton Bywater colliery, which was owned by Airedale Collieries Ltd. Allerton Main comprised four closely related pits, Albert, Victoria, Primrose Hill and Warrenhouse between them produced house, gas, manufacturing and steam coal. A considerable tonnage was shipped at Goole and Hull, but much enjoyed a wide inland market. It is known, in general terms, that certain types of coal were sent from the North Midlands and Staffordshire to the North East, where high quality household coal and steam coal was not commonly found. As T. & R.W. Bower Ltd had its registered office at Northgate, Darlington it is highly likely that these wagons supplied consumers in the Darlington and Tees-side areas. It was unusual for large operators to order wagons in batches of less than 50 and it is the author's opinion that this wagon, which was painted red, was from a batch numbered 1250-1299, built in 1930.

Introduction

Privately owned wagons, generally for the carriage of coal and other minerals, came into widespread use as early as the 1850s and their study is without doubt one of the most important and interesting facets of railway history. Unfortunately, it is also one of the most frustrating and difficult, on several main counts.

Firstly, the infrastructure, motive power, rolling stock, signalling and operating practices of the railways did not begin to change markedly (except in the case of liveries) until the mid-1960s and thus could be observed and recorded. On the other hand the glorious sight of the well maintained private owner wagon quickly came to an end, following their transfer to government control at the outbreak of World War II. While the terms of the pooling agreement did specify that wagons would be maintained and repaired by, and at the cost of, the government, the sheer horrors and operating practicalities of wartime did not allow them to be maintained as the majority of owners had done in peace time. Thus at the cessation of hostilities in 1945 much of the mineral wagon fleet was in poor condition, and much of it was life expired.

Following nationalisation of the mining industry in 1947, a considerable number of wagons were acquired by the National Coal Board for internal use at collieries. These were given a quick repaint, usually black, and in one sweep their livery was obliterated and, being on private land, they were removed from public view.

With nationalisation of the railways the remaining privately owned wagons were taken into state ownership also. Many were repainted and given a 'P' prefix to their number, but a good number, with already-fading private owner livery, simply received the new number, probably on the grounds that a full repaint was uneconomic. As the railways regenerated from the effects of the war, construction of all-steel 16 ton mineral wagons increased and wooden bodied wagons were withdrawn at an ever-increasing rate. Thus with the formation of the NCB fleet, repainting and scrapping, the sight of private owner wagons' livery had virtually disappeared by the early 1950s, long before the need to observe and record the everyday scene had become widely accepted.

Finally comes the question of records. Very many formal photographic records were taken by the railway companies and many of these survive, primarily at the National Railway Museum. Many thousands of shots of solo locomotives and three-quarter-front shots of moving trains were taken by private individuals. Conversely very few photographs of goods stock, sidings and yards were taken privately. It is true that the majority of medium and large wagon-building companies did photograph their products, but relatively little survives. Cravens of Sheffield and Birmingham Carriage & Wagon Co. records were virtually wiped out by the blitz in World War II. S.J. Claye records are thought to have been thrown away when the company was acquired by Charles Roberts, while W.H. Davies's and G.R. Turner's photographs are known to have been put on a bonfire for various reasons.

Thus the point of this introductory note is to draw attention to the fact that every piece of information regarding private owner wagons is important, and could be a missing part of the jig-saw. A portion of a wagon in the background of a photograph, a gem picked up in casual conversation with an elderly relative or retired railwayman, would be most useful. Make a note and tell your modelling colleagues. We may yet learn more about this fascinating chapter in railway history, but in the meantime I trust you will find some pleasure in the following selection of wagons, which include examples of 'a part of the story variety' that I have just referred to.

February 1996

Bill Hudson
Alfreton
Derbyshire

Plate 3: This view, *circa* 1912, has been included on account of its rarity, for the Ingleton coalfield on the western edge of the Yorkshire Dales, must have been ranked as one of the smallest, most isolated, commercially viable coalfields in Britain. Few records of the industry have survived but it is known that Hunter & Co.'s Ingleton Collieries Ltd became the New Ingleton Colliery Co. Ltd on 21st October, 1909. However, it is strange that the colliery is not listed in the *RCH Handbook of Stations, 1912*, nor in an official list of companies having sidings connected to the Midland, Lancashire & Yorkshire and London & North Western railways published in 1913. On the other hand it is recorded that the company took delivery of a Hawthorn, Leslie 0-4-0, named *King George V*, in 1910, seen here specially posed with staff and one of the company's wagons. The engine appears to be new but this must have been the condition it was kept in, for given that the wagon was repainted late in 1909, it has seen a good two years or so in service and the view probably dates from the summer of 1912. The wagon, No. 40, appears to be a standard 1907 design, with side and end doors, brakes on one side only and grease axleboxes, and is probably red, with white letters, black shading. The output from the colliery would have been consumed locally and these wagons would have travelled in north Lancashire, Westmorland and to stations on the Settle & Carlisle railway.

Chapter One

Why Private Ownership?

From the middle of World War I to the late 1920s there was considerable debate in both Parliament and the national press on the question of mineral transport and the private ownership of railway wagons. The terrible carnage of the war had resulted in serious labour shortages and the general state of the economy called for savings to be made wherever possible. It was argued that by common pooling, or outright purchase of private owner wagons by the railway companies, considerable labour savings could be achieved by a reduction in shunting and sorting empty wagons for their return journey. It was further argued that the back-loading of empty mineral wagons would be more economical. The argument continued that savings in transportation costs would be reflected in cheaper coal, both for domestic and industrial users. By the streamlining and reorganisation of wagon flows it was considered that the size of the total mineral wagon fleet could be substantially reduced, thus further reducing repair and maintenance costs and again saving labour. All these theories would be music to the ears of today's accountants, but in practice they were just not valid.

Like most of the important and valuable services in Great Britain, the system of private ownership of railway wagons was the result of individual enterprise and gradual growth. Just as the builders of highways did not provide the vehicles to be used thereon, the early railway companies did not provide wagons for the carriage of minerals, and producers, merchants and users were obliged to supply and maintain wagons themselves. This resulted in the establishment of a large number of wagon building, repairing and financing companies to meet the requirements of private owners, who, in the late 1920s, accounted for some 76 per cent of all wagons allocated to mineral transport. This system was justified by the fact that it worked well, smoothly and advantageously for the producers, distributors and consumers of minerals. Quite simply it was governed by the economic laws of supply and demand.

However there were protagonists for the abolition of the system, and its replacement by railway ownership. One of their main claims was that time and money would be saved by the common use or pooling of wagons, but in fact these two terms have completely separate meanings.

The term 'common user' can perhaps be best defined as the use of a wagon for the carriage of similar goods between convenient points. Adopted to some extent by the railway companies after Grouping, such an arrangement applied to open merchandise wagons, and covered vans. In principle sacks of imported grain could be sent from Liverpool to Sheffield, where the wagon could be loaded with boxes of nails for London. It could then transport empty beer barrels to Burton-on-Trent and so on. However wagons for any form of specialised traffic such as all classes of minerals, iron and steel, chemicals, meat, fish etc., were inevitably specially designed and were operated between specific production and distribution points, loaded out and empty home.

A degree of private common user of mineral wagons gradually developed as the 1920s progressed, for an increasing number of adjacent collieries, producing more or less similar types and grades of coal, formed themselves into loose groups or associations. The output from these collieries often had similar destinations and it was also relatively easy to draw up agreements for sharing the cost of maintenance and repair. Well known examples of such grouping include Doncaster Collieries Association (wagons lettered DCA), which comprised Brodsworth, Bullcroft, Hickleton Main, Markham Main, Firbeck and Yorkshire Main Collieries, and Welsh Associated Collieries (WAC).

The pooling of wagons, meaning the putting into a general pool of all privately owned wagons belonging to all traders throughout the country, is essentially different to the limited common user agreement referred to above. A train load of end-door-wagons with shipping coal, could be sent from Brodsworth to Hull and returned empty to Hickleton Main for a similar load. But it would be pointless to send a side-door-wagon belonging to a Hull coal merchant to a colliery for a load of shipping coal which would be unloaded by an end tippler.

On the general question of common user or pooling, railway management agreed that capital was not forthcoming for the purchase of private owner wagons; that no adequate return would be forthcoming should such capital be raised; that there would not be any reduction in the rate of hire charged for railway owned wagons; that the cost of coal would not be reduced. Furthermore the railways would only cater for normal conditions of trade, and would not provide surplus wagons as stand-bys, or for use in periods of expanding trade.

It was argued that coal wagons should be washed out and sent back to the coalfields with merchandise or agricultural produce, but in practice there simply was not the traffic. Indeed, Mr Tatlow, General Manager of the Midland Railway, giving evidence to the Coal Industry Commission presided over by Lord Sankey, then Mr Justice Sankey, on 7th March, 1919, said,

Taking the Midland Railway as an example, that railway conveys to London from the colliery districts between four and five million tons of coal per annum. None of these wagons (even if they belonged to the railway) are required for loading goods out of London, because our experience is that, dealing with merchandise traffic alone, the number of loaded wagons worked into London is greater than the number of wagons required to load traffic out of London. So that, leaving the coal wagons out of the question, we have to work empty goods wagons from London there being no freight to put in them.

In statistical terms it could be put that miscellaneous non-mineral merchandise traffic amounted to less than 20 per cent of the total quantity carried. If mine workers and their dependants were estimated at one-tenth of the population, it would follow that those persons would call for return loads of less than 2 per cent of the total merchandise traffic of the country. Thus, in 1928, the mining districts sent out 248,903,000 tons of coal and received, say, one-tenth of the 57,226,000 tons of general merchandise; that is about 2½ tons inwards for every 100 tons outward.

During World War I regulations under the Defence of the Realm Act authorised the use of wagons returning empty, but on the few occasions this power was used the end result was only to cause delay and inconvenience to the collieries awaiting the return of their empty wagons. The railway companies were well aware that instead of wasting time and money on providing washing facilities and drainage sidings to clean out coal wagons, it paid best to send them straight back to be re-loaded. These practical considerations dispelled the doctrinaire notion that any acquisition by railway companies, or a general pooling of privately owned wagons, would materially affect return haulage.

In evidence given before the Royal Commission on the Coal Industry, 1925, it was estimated that if private owner wagons were railway owned, a saving might be effected of one-eighth of the time spent in shunting. Sir Ralph Wedgwood, then Chairman of the Railway General Manager's Conference could not see that more than one penny per ton would be saved by such diminished shunting, while the Mining Association of Great Britain put the maximum saving nearer one farthing per ton. All railway estimates in regard to shunting time dealt solely with tonnage quantities, utterly disregarding the kinds of traffic involved. Yet there were more than 50 grades of coal carried by the railways and receiving facilities differed greatly at the terminals. With such a diversity of types of coal there were many types of wagons with end doors only, side doors only, end and side doors, side and bottom doors, and so on. Each type of wagon was designed for a purpose, be it shipment, gas works, industrial coal, blast furnaces, household coal or otherwise. The collieries knew precisely what type of wagon each customer required and, provided it had its own fleet of empty wagons to hand, it could shunt them internally as required. Supposing that on a given day the colliery wanted wagons for supplying 20 different kinds of coal to 10 classes of customers, what kind of shunting task would have to be performed at the railway marshalling yards if all wagons were railway owned and identified only by a running number? It would be virtually impossible.

Collieries, generally speaking, were not designed so that coal raised day by day was stored on the ground. It was universal practice to take coal to the screens and washeries, from where it was loaded direct into railway wagons. The raising and dispatch of coal was always recognised as a continuous process and if at the beginning of the day there were insufficient empty wagons to hand the colliery would not turn coal. It was with a view to ensuring a regular, sufficient and adequate supply of wagons that the very system of private ownership came into existence in the mid-19th century. Every private owner kept a staff of clerks to check the whereabouts of his wagons. There were wagon chasers at junctions and yards whose duties were to see that empty wagons were speedily returned. The capital expended in establishing or developing a colliery could lie idle for long periods if reliance had to be placed on there being a sufficient supply of railway owned wagons. After all the railway companies were not under any statutory obligation to ensure that sufficient wagons were available day by day. The Railway Traffic Acts only made them responsible to afford facilities 'according to their powers', while the

Court of Railway Commission could not oblige them to raise capital for new enterprises.

If all wagons were owned by railway companies, the producers and distributors of coal and other goods would have been entirely at the mercy of the railway companies, without any enforceable claim for redress. It was argued, quite rightly, that it would be neither just nor desirable to place the producers, distributors and customers wholly under the control of the railway companies.

In the valleys of South Wales, and in many other places, space was not available for the storage of coal in colliery yards. If it was placed on the ground and had to be lifted by hand or machine into wagons, disproportionate cost would arise, and the value of the coal would be depreciated by breakage. With the softer kinds of coal used for industrial and household purposes there was also the danger of spontaneous ignition. Thus on economic and safety grounds it was customary to load coal straight into wagons. If inclement weather or periods of slack trade did not allow the immediate dispatch of coal, an adequate supply of wagons allowed them to be used for storage and the colliery could continue to raise coal. In this respect more modern collieries were laid out with sidings to accommodate five or six days' production. At blast furnaces, iron works and other large places of custom it was common practice to keep coal in wagons until it could go direct into the furnaces or boilers. It was not unusual to find up to 500 wagons standing for this purpose. In addition, large coal factors such as Wm Cory or Stephenson Clark would often keep coal in wagons at junctions and yards ready for immediate dispatch to customers.

These are but typical examples of the use of wagons for storage, and while so doing the private owner did not have to pay the railway companies a wagon demurrage charge. If, on the other hand, the wagons were railway owned, the companies would certainly levy a charge while the wagons were being used for storage. Additionally it would probably have been the practice of the railway company to ascertain how many wagons each producer would require to deal with normal output and to allocate to that producer his apportioned quota. If he applied for more, when it was known that he was using them for storage, or indeed he needed more because of exceptionally high demand, he would no doubt have been refused on the grounds that he had exceeded his quota. Thus he would have to stop raising and employees, machinery and capital would stand idle.

In the North East railway wagons were only supplied sufficient to deal with the tonnage of coal to be dispatched each day, and collieries in Northumberland and Durham had to be content with the limitations to which their systems confined them. Most of their owners readily admitted they would have been glad to have the facilities enjoyed by private owners.

If all wagons were pooled immense and practically insuperable difficulties would arise with regard to repairs, owing to differences of construction, the varying standards of maintenance and the number and variety of spare parts that would have to be kept at depots, to say nothing of the complexity in apportioning the share of costs of repairs. All loaded wagons sent out were followed up day by day by the owner, but had they been pooled no one would have had a particular interest in any specific wagon. The old money-saving adage that 'a stitch in time saves nine' would have ceased to apply. Under the private owner system, wagons needing repair were made good by the owners or by wagon repairing companies employed by them, and competition between repairers kept costs to a minimum. When arrangements were made for repairs the parties knew approximately the amount of wear and tear the wagons were likely to sustain and could contract accordingly. However, running conditions under a pooling system would have been an unknown quantity, and no equitable system would have been evolved whereby persons using pooled wagons could have paid for repairs proportional to their use of the wagons.

Under the existing system the majority of private owner wagons ran more or less on regular paths and in districts where repair parts were readily available. If, however, a wagon domiciled in South Wales or East Anglia had been sent to the North of Scotland and broken down, a long and costly delay of up to 10 days could have ensued while parts were ordered and sent north. In fact during World War I thousands of wagons were sent to the far north with coal for the warships based at Scapa Flow. Under the prevailing conditions any wagons which broke down were virtually abandoned. It was reported to the Sankey Commission in 1925 that the inability to get wagons repaired during the war virtually brought the Great Western Railway to a standstill.

There were at least 150 companies involved in wagon building (although some were little more than two man businesses), and a considerable number in wagon repairing, employing an estimated 10,000 persons. If private building of wagons had ceased, those workers and their dependants would more than likely have been forced to move to find alternative work, creating economic havoc in the communities where they lived and worked. In the 1920s there was not the welfare system we take for granted today.

In the late 1920s there were an estimated 600,000 private owner wagons, about a quarter of which had been built in the previous 16 years or so. At a conservative estimate those wagons were worth £50 million,the bulk of which had been provided by colliery proprietors so as to ensure full and regular working of their mines. It would have been a surprising, and foolish, venture, for railway companies to have tried to raise the capital to purchase these wagons at a time when motor vehicles were already becoming formidable competitors to the railways. Private wagon owners were financed by money lent on debentures by a large number of investors. If the system of private ownership had been abolished, it is doubtful that those investors would have put their money into ordinary stock of the railway companies. The system of private ownership was of great value to colliery companies in purely financial terms, for it allowed them in both difficult and expensive times to obtain circulating capital, simply by pledging their wagons on terms which allowed them to repay the principal by easy installments.

The existing system was also of great financial benefit to the railway companies, as wagons were provided by private owners and the building, repairing and financing companies. These wagons were available to the railway companies as means by which charges for carriage could be earned, but upon which no capital expenditure whatsoever had been incurred.

Bearing in mind all the points raised in the foregoing, it is not surprising that the Royal Commission on the Coal Industry unanimously reported, in March 1926, that 'that they could not recommend that compulsion should be applied as regards the ownership or pooling of private wagons.' Furthermore they found that all the parties directly affected, consisting of railway companies, the mine owners, and, the traders were definitely opposed to private wagons being brought into ownership of the railway companies.

Thus the system of private wagon ownership remained in force until the dark days of 1939, when the outbreak of World War II overshadowed all other considerations. With the prospects of a long and bitter conflict ahead the provision of one unified transport system became of paramount importance and, together with all the wagons not built for special use, virtually all private owner wagons were thrown into the general pool under the control of the government. After cessation of hostilities, the railways remained under interim government control for a while, but a mere six months after the end of the war the newly elected Labour government announced its intention to nationalise the railways. As a result private wagons were never returned to their owners. Those wagons still running in the pool were purchased by British Railways, but with increasing construction of the 16 ton all-steel mineral wagon introduced during the war, they were rapidly withdrawn and scrapped, with a final clear-out in 1958/59.

For a considerable period after Nationalisation all wagons were owned by BR, but the principle of private ownership would not go away and by the early 1980s PO wagons, owned primarily by the roadstone and aggregate producers, were once again on the rails of Britain. Today the wheel has turned full circle and with privatisation of the railways all wagons will be 'private owner'. The ghosts of the colliery owners, merchants and dealers who fought so long to keep their wagons, must indeed wear a wry smile!

Chapter Two

Colliery Owned Wagons

Throughout their history approximately 70-75 per cent of private owner wagons were operated by the colliery companies. The earliest wagons, of the chaldron type, carried 4 tons, but 7 and 8 ton wagons had become common by the 1860s, with the 10 ton wagon being introduced shortly afterwards. By the early years of the 20th century the 12 ton wagon had been developed, and although the 15 and 20 ton wagon types were to appear later, the 10 and 12 ton vehicles were to remain by far the commonest in use. By the 1930s the depression in the general economic climate, and the mining industry in particular, had seen a fall in coal production from 287 million tons (in the record year of 1913), to 228 million tons (in 1936). Nevertheless, the coal owners still operated a fleet of some 324,000 wagons, or roughly 3½ times as many as the coal factors, distributors and merchants. Thus, with the exception of Northumberland and Durham, where almost all the coal output was carried in railway company wagons, colliery-owned wagons could be seen everywhere, even down to the humble West Country or East Anglian branch, and their importance to the historian and serious railway modeller cannot be overestimated.

The wooden mineral wagon was, in simple terms, a number of planks or sheets, held together by various metal straps and carried on a substantial wooden frame designed to absorb the longitudinal forces resulting from haulage and shunting. Although there were, of course, many detailed differences, the two basic types of private owner coal wagon, of interest to the majority of modellers, were the Railway Clearing House (RCH) designs of 1907 and 1923, exemplified by *Plates 1 & 2*.

Although the author may appear to imply that a whole new specification was prepared, this is far from true, particularly with regard to the 1907 wagon. At the turn of the century there were over 150 firms engaged in wagon building, but only a handful of these were in the 'big league', and very near, or actually at, the top was Charles Roberts and Co. of Wakefield. As such the company had considerable influence, and when the RCH decided to adopt a standard wagon design, to all intents and purposes it simply adopted what Roberts & Co. had been building for some years.

While not mandatory, for the Gloucester Railway Carriage & Wagon Co., for example, carried on regardless with its own designs, the specifications were widely adopted, not least as a cost effective measure. The standardisation of parts not only allowed mass production in the first place, thus reducing cost, but also permitted the interchange of parts between various manufacturers. Bearing in mind the huge number of builders' 'out stations', usually an 8 ft x 12 ft hut, and the number of small independent wagon repairers, many situated at remote yards and junctions, this was indeed a useful facility.

By the time the 1923 wagon was introduced, shortly after Grouping, the concept of centralised control had become much stronger and the specification became mandatory for the construction of new wagons. Not only did private builders turn to this design, but it was also adopted as the standard mineral wagon by the LMS and LNER. Apart from differences in doors the body could be built in 7- or 8-plank configuration. The higher-sided wagons are said to have been used because it was not possible to get 12 tons of Yorkshire coal in the 4 ft 4½ in. wagon.

Plate 4: Under the RCH 1907 regulations some latitude in body design was allowed, as illustrated by this 10 ton wagon, which has 3 x 9 in., 1 x 8 in. and 2 x 7 in. planks. It is fitted with side, end and bottom doors, lipped buffers at the door end, brakes on one side only and Attock's 46 axleboxes. It is painted best red (London red with a trace of chrome yellow), with white letters, black shading, and is from a batch of 50 wagons, ordered on 28th January, 1908 and numbered 301-350. A further batch, numbered 351-400 followed in February, and on 25th May, 1908 a batch of 100, Nos. 701-800 were ordered. The italic lettering reads *'Empty to Brodsworth Main Colliery Sidings, H&B Ry.'* The colliery was connected to the Hull and Barnsley Wrangbrook Jn-Denaby line, and the Great Northern Doncaster-Leeds line, and produced high quality steam, household and gas coal from the famous Barnsley Bed. The hard steam coal was in great demand by shipping companies and something like 50 per cent of the annual output of 1.5 million tons was bunkered at Hull, Grimsby, Immingham, Goole, Keadby, Liverpool and Manchester. Inland, the quality of the coal took the wagons virtually everywhere.

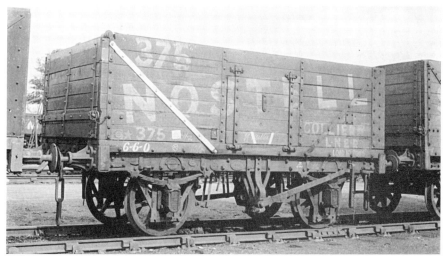

Plate 5: This view, taken about 1944/45 illustrates the longevity of the 1907 wagon. It has side, end and bottom doors, Attock's 46 axleboxes and retains 'handed' brake shoes (i.e. right- and left-hand versions with fixing lug at top only). It is almost as built but it has received the second set of brakes, ribbed buffers and standard side door bang stop. These wagons were originally red, with white letters and black shading, but during the 1930s this was changed to the cheaper black, with unshaded lettering. Nostell colliery, which produced manufacturing, house and steam coal, was connected to the GNR Doncaster-Leeds line, and while coal was shipped at Hull, Goole, Grimsby and Immingham, much was consumed in East Anglia and the South East.

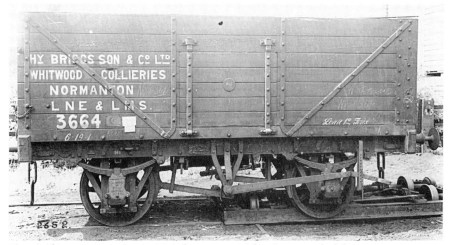

Plate 6: This rare view, taken in 1934, shows a second-hand wagon which has come in for repair and overhaul. Originally built for Doncaster Collieries Association in August 1922, it is fitted with side, end and bottom doors, lipped buffers at the door end, brakes on either side and oil 116 axleboxes. When photographed it was black, with simplified lettering in unshaded white, which may well have been applied at one of the pits. When it emerged after overhaul it probably carried the usual Briggs' livery of black, unshaded name in 30 in. letters right across the side. The company had five collieries, Whitwood at Normanton (MR & NER), Haigh Moor at Robin Hood (MR & GNR), Water Haigh at Woodlesford (MR), Saville at Methley (MR) and Snydale at Featherstone (L&YR). While some 80 or so wagon loads were shipped daily at Hull, Goole and Liverpool, inland markets included the North West, Yorkshire, Lincolnshire, the Midlands, East Anglia and the South East.

Plate 7: South Kirkby, Featherstone and Hemsworth Collieries Ltd was one of the largest colliery undertakings in Yorkshire, with an output of almost two million tons in 1913, and a wagon fleet in excess of 2,500. The three collieries produced house, gas, manufacturing and steam coal and some 20 per cent of output was shipped at Hull, Goole, Immingham and Grimsby. Inland wagons would have travelled to Lancashire and the North West, Yorkshire, the Midlands and much of eastern and southern England. Each colliery had its own wagons, with separate liveries, but they were interchangeable as, the italic lettering on the wagon illustrated reads *'When empty return to Featherstone Sidings, L&Y Rly.'* Of the three, South Kirkby had the largest fleet, Charles Roberts building them 1,420 wagons between 1899 and 1916. The wagon shown is from a batch of 100, numbered 2201-2300, ordered in April 1911 and fitted with side end and bottom doors, Ellis 12 ton axleboxes, brakes on one side only and side door springs. They had 5 x 7 in., 1 x 8 in. and 1 x 9 in. planks and were painted grey (same as LMS grey), with white, unshaded letters. The small block lettering reads 'GN. GC. NE. & MIDLAND RAIL$^{\underline{YS}}$'. Between the stanchions on the fixed end was lettered:

Owner
John Shaw
2201 (etc.)

PEA SLACK.

FROM

SOUTH KIRKBY COLLIERIES

West Riding and Grimsby Joint Line (G. N. & G. C. Rlys.) and
Swinton and Knottingley Joint Line (Mid. & N. E. Rlys.)

To _Irlam (GC)_____ Station,

Via _____

For M _W. Taylor_____

	TONS.	CWTS.
Wagon No. 754 : Weight	8	18

Date 10 - 4 - 1902

A typical South Kirkby wagon label, showing 8 tons 18 cwt of pea slack (½ in. x ¼ in.) sent via the GC route to Irlam, for Mr W. Taylor on 10th April, 1902. Being small coal this was probably for use in an industrial boiler or furnace.

Plate 8: The Stanton Iron Company owned collieries to supply both its own works and the open market. The pits at Pleasley, Silverhill and Teversal, which were connected to the MR, GC, GN and LDEC routes, were producing over one million tons per annum of house, manufacturing and steam coal, even before Bilsthorpe was sunk in the late 1920s. This enjoyed a very wide inland market and was also shipped at Boston, Goole, Grimsby, Hull, Immingham, Kings Lynn and Liverpool. The wagon illustrated was from a batch of 100, numbered 3348-3447, built by S.J. Claye and registered by the LMS in February 1924. It was fitted with side, end and bottom doors, brakes on either side, ribbed buffers, pin and chain end door fastening and united Ellis axleboxes. The side sheeting is unusual and appears to comprise 2 x 7 in., 1 x 5 in., 2 x 7 in. and 3 x 6 in. planks. It was painted red, with white letters and black shading. The italic lettering reads '*Empty to Teversal or Pleasley Collieries*', below which is the tare of 7-1-1.

Plate 9: Welbeck colliery near Warsop, Notts, was sunk in the 1920s by the New Hucknall Colliery Co. and produced house, steam and manufacturing coal. The colliery had access to the GC, GN, LDECR routes and the MR via Shirebrook, and had wide inland markets in Lancashire (via Rowsley), the Midlands and the South. Coal was also shipped at Boston and the Humber ports. The wagon illustrated is a 1923 standard wagon, built by W.H. Davis, and has been included on account of its unusual livery of black body and unshaded yellow lettering.

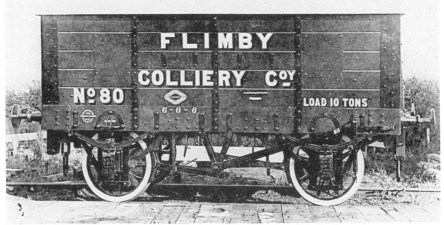

Plate 10: A characteristic of the Cumbrian mining industry was the widespread use of hopper wagons, illustrated by this example built by R.Y. Pickering in October 1908 and registered by the LNWR. It appears to be 16 ft 0 in. x 7 ft 6 in., with 5 x 9 in. planks, and is fitted with bottom doors, brakes on one side only and round bottom axleboxes. The top plank is cut back to allow for a 2 in. x 2 in. outward-facing angle-iron capping, flush with the body side, to protect the woodwork. It also carries the heavy duty end stanchions, extending below the headstock to buffer up to chaldron style wagons. The wagon is painted red, with white letters and black shading. Flimby Colliery Co. had three pits, Bertha, Robin Hood and Watergate, situated between Flimby and Maryport. Some coal may have been sent to Northern Ireland, but most would have been consumed by local industry, perhaps as far south as Millom and Barrow-in-Furness.

Plate 11: Although apparently photographed as new in July 1898, this wagon was built the previous year. The most likely explaination is that it was built speculatively, registered with the Gloucester Co. and photographed after lettering and hiring to Camerton Collieries. It is a typical Gloucester 10 ton wagon, with side and end doors, brakes on one side only, Ellis patent axleboxes, V-shaped vee hangers and internal diagonal side braces. It measures 14 ft 11 in. x 7 ft 5 in. with 1 x 5 in. and 6 x 7 in. planks and is painted black, with unshaded white letters. The italic lettering reads '*Empty via Hallatrow*'. It was thought that the markets of Somerset collieries were restricted to southern and south western counties, but it is now known that Camerton

Collieries supplied coal to Keswick, presumably for the gas works. (As an aside it has also come to light that Scottish PO wagons ran to Cumbria.)

Plate 12: A peculiar quirk of many Welsh colliery owners was the desire to let their wagon liveries take the form of a short essay, as this superb example of the signwriter's art shows. Built in 1910, and registered by the GWR, this 12 ton wagon is fitted with side and end doors, brakes on one side only, V-shaped vee hanger, wooden door stop opposite brake and Gloucester No. 7 axleboxes. It has two commode handles at each end, a requirement for wagons working to South Wales ports, and buffers with foottreads. The wagon is 16 ft 6 in. x 7 ft 10 in. x 4 ft 2 in. with 5 x 7 in. and 2 x 9 in. planks and is painted black, with unshaded white letters. The italic lettering reads '*Empty to Onllwyn, Neath & Brecon Rly*'. Anthracite was considered to be the finest coal in the world and consisted of up to 94 per cent carbon. Due to its hardness, once washed, it gave off no dust and was thus clean to handle. It burnt very slowly and evenly giving economical, easily regulated even heat. It was almost smokeless and left practically no ash, making it the fuel *par excellence* for central heating, domestic boilers, green houses, brewing, malting etc. It was ideally suited for use in suction gas plants in the making of producer-gas which was widely used in heavy industry, particularly the steel industry. Anthracite is only found in the northern part of what were Glamorgan and Carmarthen, in South Wales, and to a lesser extent in the Central and Fife and Clackmannan coalfields, in Scotland.

In 1930 the output of anthracite from South Wales was around six million tons, half of which was exported. Inland the wagons would have been seen virtually anywhere, especially the hop drying areas of Kent, the market gardening areas of central and eastern England, the steel producing areas of the north, towns with breweries, more affluent towns where centrally heated houses had a significant presence and so on.

As the demand for anthracite grew its production came under far greater centralised control than any other part of the mining industry. In 1923 the four collieries of the Cleeves Western Valleys group joined with six others to become Amalgamated Anthracite Collieries. Soon afterwards another 10 combined to form United Anthracite Collieries. In 1927 these two groups united to form Amalgamated Anthracite Collieries Limited. Thereupon the constituent companies were liquidated, the combined group becoming one operating group, with a common wagon livery. Only a year later another round of acquisitions took the holding up to 34 collieries, with a wagon fleet of 15,000. Although one basic wagon livery was adopted the author is aware of at least six variations, as shown next.

Plate 13: This view at Upper Bank, Swansea, in 1938, consists of 10 ton side and end door wagons, with minor variations, but it is the liveries which are of greater interest. The near wagon, No. 20573 (square bottom grease axleboxes, side door spring) has 'ANTHRACITE' on the middle three planks, 'A.A.C.' above the door, and 'SWANSEA' in 3½ in. letters at the top of the bottom plank at the right-hand end. Above this in italics is 'Amalgamated Anthracite Colls Ltd.' Above the number, in two lines of italics is '*Empty to Felin Fran, GWR S. Wales*'. '*Load and tare (6-7-0)*'.

The second wagon, No. 14366 (Ellis patent axleboxes, wooden door stops) has the same initials and main name, but 'SWANSEA' in 5 in. letters at the bottom of the plank. Small italics as last, but also in two lines on bottom plank of door: 'Repairs Advise Owners'. © mark to the left of number, 'Load and Tare (6-3-0)' in block letters. Next, No. 36293 (no door stops, Gloucester No. 4 axleboxes) has 'A.A.C.' the same, but 'ANTHRACITE' in 10 in. letters on the fourth and fifth planks. Different, but illegible italics on door, 'SWANSEA' obscured by number taker. © mark and yellow star below 'A'

of 'ANTHRACITE', 'Load and tare (6-7-0)' in block letters, No. on top of plank at end. The fourth wagon, No. 7520 (side door spring, Attock's patent axleboxes) has 'ANTHRACITE' in 12 in. letters at top of the fourth and fifth planks, with 'AMALGAMATED' in an arc over the middle four letters of 'Anthracite'. Below the main name on part of the second and third planks, in 9 in. block letters is 'COLLIERIES LTD SWANSEA'. On bottom plank at right in capitals is 'AND' (5 in.) 'LONDON' ('L' at 7 in., rest at 6 in.). To the left of metal number plate, on two lines, 'Felin Fran, ?' Above this in large, crude italics, '*Empty to*'. These four wagons are all black, with unshaded white letters.

On the third row, behind the Brynhenllysk Colliery Co. wagons, is a wagon with 'AMALGAMATED' in an arc; but painted light grey, with white letters, black shading. Finally, further along the same row is an A.A.C. wagon, with the first 'A' and the 'C' outside the side knee washer plates (the vertical ironwork either side of the door.)

Chapter Three

Distribution

The coal distributor had three prime tasks, to find a market for the various grades of coal produced at a colliery; then persuade the consumers to buy these coals regularly; and finally to distribute this coal with efficiency, economy and progress. The consumer also looked to the distributor to find the right coal for his particular requirements. When it is realised that there were some thousands of collieries, each producing coal of varying sizes, from different seams, it will be seen that the perfect distributor would have to have been a walking encyclopedia. In reality distributors specialised in a few branches of the trade and their managers were able to acquire an intimate knowledge of the coals they sold.

The various branches of coal distribution may be put into four classes, the exporter, the wholesale merchant or factor, the merchant and finally, the dealer. These demarcations were not always rigid, however, for the big names in the game, such as Wm Cory, Charringtons, Stephenson Clark, Ricketts etc. performed all the classes except for that of dealer.

The exporter ships coal in large quantities to foreign buyers, to his own depots in foreign countries, or to his bunkering depots at various points on the world shipping routes. He may have used his own wagons between the colliery and the port or he may have arranged for the colliery to deliver so many thousand tons to a given port at a given time.

In 1930, some 70-80 million tons, or roughly half the amount distributed for inland consumption, was handled by the wholesaler. The wholesale merchant bought very large quantities of coal of all qualities and descriptions, and supplied the ships, barges and railway wagons in which to transport it. It is a well known fact that coal consumption is seasonal and this could have a potentially disastrous impact on the operation and income of the collieries. However by the size of his operations, the diversity of his markets and his ability to store coal, the wholesaler could more or less even out this fluctuation in demand and take regular supplies from the colliery. Conversely, he was able to overcome that equally disastrous problem, where a large industrial consumer was faced with an unexpected demand for fuel which could not be met out of his own normal stocks.

A valuable service rendered by the wholesaler to the colliery was the purchase of large quantities of coal before it was needed. With large overhead costs, and responsibility for the wages of thousands of workers, a colliery could not function without an assured market for the greater proportion of its output for a considerable period ahead. There is another branch of wholesale trading known as factoring, whereby the wholesaler supplied coal to the small merchant who was not in direct touch with the colliery. This provided a further market for the colliery and removed the need for the small trader to attend the periodic coal markets. It also helped the merchant in that he knew the factor would use his expertise to supply the right coal for his particular needs.

The coal merchant was the holder of a depot which received rail-borne supplies from the colliery and from which road deliveries were made to the domestic and small industrial consumers. It was the duty of the merchant to store coal during the summer and by so doing he helped to keep the pits at work and create a reserve against those days in winter when demand exceeded supply.

It was this uneven demand between winter and summer which was the biggest worry for the merchant. He could not afford to maintain a stud of horses or staff of clerks all the year round to cope with the demand of four or five months. Yet, if in a period of prolonged cold, his organisation became strained, he was accused of inefficiency and lost trade. As the 1920s and 30s passed the life of the merchant became tougher through no fault of his own. In earlier times houses were built with capacious cellars, and the merchant kept his plant employed filling these at his lowest summer prices. In modern, smaller houses an outside coal store of less than half a ton capacity was considered adequate and every cold snap brought a flood of small orders, all marked for immediate delivery. Under the circumstances the cost of distribution rose and the merchant was criticised, yet it was the architects who were far more to blame than the merchant himself.

Finally at the bottom end of the coal distribution network sat the dealer. He bought coal from the merchant's wharf and literally hawked it round the streets, selling to small households, to those who could only afford a little at a time and to the poor through coal shops. He never rose to the dizzy heights of owning his own wagons and may have had no more than a handcart.

Plate 14: J.H. Beattie & Co. Ltd were wholesalers and contractors to HM Government. They would deliver truck loads to any stations in the south of England, or by road to any part of London, from one of their four depots. These were at Somers Town and Pancras Road Bays (MR), Brompton & Fulham and Queens Park (LNWR). Much of their coal was obtained from Derbyshire and Nottinghamshire, reaching the MR depots via Toton and Brent, and the LNWR via Wigston, Nuneaton and Willesden. It is also known that in 1939

they supplied anthracite nuts from Gurnos colliery, Glamorgan, to Wimbledon Corporation Electricity works. The wagon illustrated is one of a batch built by Chorley Railway Wagon Co. very early in 1923, and registered by the LMS. It is fitted with end, bottom and side doors with folding top doors and Attock's 158a pattern axleboxes. The buffers at the brake lever corners have foottreads, and above these commode handles are fitted across the second and third planks. The wagon was 16 ft 5 in. x 7 ft 11 in. with 5 x 7 in. and 1 x 9 in. planks and was painted black, with unshaded white letters.

Plate 15: The oldest firm in the London coal trade, founded in 1731 by Joseph Wright, Charringtons, by a series of amalgamations eventually became Charrington, Gardner, Locket & Co. Ltd in 1922. Described as coal and coke contractors and retailers they had, in 1930, a fleet of almost 3,000 wagons. The company was a contractor to many power stations and large industrial concerns, but was primarily interested in the domestic coal trade, having over 50 depots in London alone. These included Somers Town, Cambridge St, Finchley Road (MR), Finchley Rd & Frognall, Gospel Oak (LNWR), Brixton, Brockley Lane (SECR), Kensington High St (Met.), Old Ford (NLR), Cable St (GER) and Kew Bridge (LNWR/MR/NLR Jt). The firm obtained coal from all the major coalfields and its wagons could be seen virtually anywhere. The vehicle illustrated is a standard 1923 RCH design, built in February 1934 by the Derbyshire Carriage & Wagon Co. However, it had top doors, in addition to the side, end and bottom doors and the sheeting was unusually arranged as 4 x 7 in. and 3 x 9 in. planks. It also had two commode handles on the third plank of the end door. It was painted red, with white letters and black shading. the italic lettering reads '*Charrington, Gardner, Locket & Co Ltd*
16, Mark Lane, London, EC3'

The plate next to the label clip reads:

	Empty to	
Toton Sidings	LM&SR
Colwick Sidings	LM&SR & LNER
Whitemoor Sidings	LNER
Neasden Sidings	LNER

Unless otherwise labelled

Plate 16: Originally set up as a private company in 1787, Wm Cory & Son Ltd came into being in 1896 when seven other companies were absorbed. By 1930 it had become the world's largest coal distributor responsible for the wholesale handling of 20 million tons per annum. It supplied coal for all industrial purposes, electricity and gas making, pulverisation and domestic uses, and was contractor to the Admiralty and all the leading steamship lines. The wagon illustrated, built by Hall Lewis very early in 1924, must have been one of the last built to a pre-1923 specification. It is fitted with side and end doors, united Ellis axleboxes and plain buffer guides with foottreads. It has two large commode handles and very distinctive D-dropper end door catches. It is 16 ft 6 in. x 8 ft 0 in. with 1 x 5 in., 5 x 7 in. and 1 x 9 in. planks and is painted red, with white letters and black shading. Italic lettering reads *'Empty to Toton Sidings, MR.'* This livery seems to have disappeared in the early 1930s.

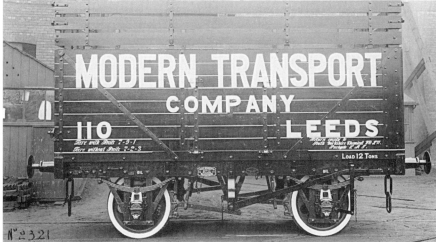

Plate 17: The Modern Transport Co., which was sales agent for Earl Fitzwilliam's Elsecar Collieries, was a substantial shipper, distributor and retailer with depots in London, Liverpool, Leeds, Hull and Grimsby, and it is believed the firm had wagons lettered for each of these depots (see the author's, *Private Owner Wagons, Volume Three*, plate 10, Oxford Publishing Co.). The wagon illustrated is a 1923 RCH wagon, fitted with side, end and bottom doors and removable coke rails. It is one of a batch built in November 1931, numbered 100-129 and painted black, with unshaded white letters.

Italics read: *Tare with Rails*   7-9-1   *Return Empty to*
               Tare without Rails   7-2-3   *South Yorkshire Chemical Wks Ltd*
                                                 Parkgate LMS

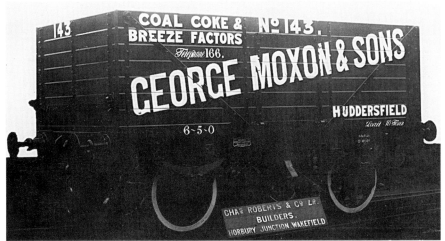

Plate 18: G. Moxon & Sons supplied coal, coke and breeze (small, broken coke) to small industrial firms and coal merchants in the Calder and Colne valleys, and on the L&YR line to Penistone and Barnsley. The wagon Illustrated is a standard 1907 wagon, from a batch of 10, numbered 135-144, ordered on 13th October, 1908. It is fitted with side, end and bottom doors, side door stops and Attock's 46 axleboxes. It was followed three weeks later by a further 10, Nos. 145-154. For painting the order book gives slate, vermillion band corner to corner, white letters, all shaded. Traces of the band can be picked out on the original print but the film emulsion in use at the time did not pick up red very clearly.

Plate 19: To make best use of their vehicles at a time when the demand for coal was at its lowest some of the traders were also builders' merchants, and during the summer their wagons would carry bricks, earthenware pipes, tiles, slates, sand, gravel etc., giving them a wide area of operation. Messrs Parker & Probert, Birmingham, operated such a business in the early years of the century. The wagon illustrated was built in 1898 and is fitted with side doors only and No. 4 axleboxes. It is 14 ft 11 in. x 7 ft 5 in., with 7 x 7 in. planks and is painted chocolate, with white letters and black shading. The italics read '*Empty to Brereton Collieries, Rugeley, LNWR*'. In 1900 the company had depots at Wellington Wharf, Bagot Street, Witton Wharf, Aston Lane and Soho Pool wharf. In 1911 it was listed at Witton Wharf and Erdington. The following year it was not listed and is believed to have split up.

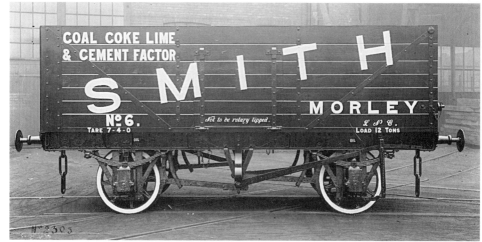

Plate 20: This plate shows a quite mysterious wagon operated by A.H. Smith, Morley. Photographed after overhaul in 1929, the wagon is obviously second-hand and may well have been built for Government use overseas during World War I. It appears to be 18 ft 0 in. x 7 ft 11 in. with a 10 ft 6 in. wheelbase and 5 x 7 in. and 2 x 9 in. planks. It is fitted with side, end and bottom doors and Wagon Repairs Ltd oil axleboxes. It is painted black, with unshaded white letters. The company also operated three conventional 12 ton wagons, Nos. 10-12, with an identical livery, but painted red, with white letters and black shading. Coal and coke would have come from south and west Yorkshire and lime from Womersley, Settle and possibly Derbyshire.

Plate 21: Seen here is a merchant's wagon from Rock Ferry on the Birkenhead Joint Line (LNWR & GW). It was built by Hurst, Nelson, Motherwell about 1907, and is fitted with side doors only and Rigley's patent axleboxes. An unusual feature is the two short steel door stops arranged to coincide with the hinges. The wagon is 16 ft 0 in. x 7 ft 6 in., 2 x 7 in., 1 x 5 in. and 4 x 7 in. planks and appears to be painted grey with white letters, and black shading. Coal would have been obtained from North Wales, North Staffordshire, and Notts/Derbyshire. By 1926 Joseph's son had taken over and the wagon would have been lettered 'JOHN DAVIES'.

Plate 22: Although of poor quality this view has been included for illustrations of wagons from Central Wales are rare indeed. There is nothing to indicate the builder, but it is an 8 ton wagon, with side doors only, with wooden door stops, brakes on one side only and round bottom grease axleboxes. The low tare weight (4-18-0), the buffers and the diamond plate on the solebar, indicate that the wagon has been converted from dumb buffers. As such it is probably 14 ft 6 in. x 7 ft 6 in. and it has 3 x 9 in., 1 x 5 in. and 1 x 9 in. planks. It appears to be painted red, with white letters and black shading. Newtown lies on the Cambrian line between Welshpool and Moat Lane and supplies were probably obtained from South Wales, Wrexham, and Cannock Chase and South Staffordshire.

Plate 23: This view shows an 8 ton wagon built in April 1898. It is a standard Gloucester wagon, with side doors only and No. 4 axleboxes. It is 14 ft 11 in. x 7 ft 6 in., with 4 x 7 in. and 1 x 9 in. planks and is painted red, with white letters and black shading. Andoversford lies on the GWR Cheltenham to Stow-on-the-Wold line and coal would have come from the Forest of Dean or Warwickshire and Notts/Derbyshire via Banbury and Rugby. As in *Plate 20*, by 1926 the son had taken over and the title changed to 'W.J. FINCH'.

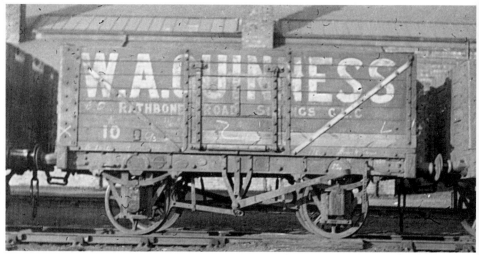

Plate 24: At the end of the war the PO fleet was more or less still intact, but in a sorry state, as this *c.* 1945 view of a standard 10 ton wagon shows. Believed to have come from the Chorley Wagon Co. (plate just visible to right of V-hanger), it is fitted with side and end doors and very plain grease axleboxes, on the visible side, one round and one square bottom. It was 16 ft 0 in. x 7 ft 6 in. with 7 x 7 in. planks, and under the grime it was probably painted red, with white letters and black shading. Still visible is the white painted diagonal brace, applied to PO wagons after they came under Government control in 1939. Strangely the wagon is lettered 'Rathbone Road Sidings C.L.C.', but these were at the end of a LNWR branch from Edge Hill, Liverpool. W.A. Guinness would have supplied the small industrial and the domestic market with coal from North Staffs, Lancashire and the Yorkshire/Derbyshire/Notts area.

Plate 25: It is surprising how far the sales reps of the big wagon builders travelled, as this example of a 10 ton wagon built about 1905 by Hurst, Nelson, Motherwell for R. Hill & Son, of Harwich, shows. It is fitted with side doors only, with lift-over top planks, brakes on one side only and Rigley's patent axleboxes. The two wooden door stops are worthy of note. The wagon appears to be 15 ft 6 in. x 6 ft 11 in., with 1 x 5 in. and 6 x 7 in. planks, and is probably red, with white letters and black shading. It is unusual to see the main name carried right to the top of the wagon and to see shading overlap onto the plank below. This merchant would have obtained house coal from Notts/Derbyshire via Colwick, Peterborough, March and Cambridge and from Yorkshire via Doncaster and March.

Plate 26: This is one of the few known illustrations of work by W.H. Davis & Son of Langwith Junction, Mansfield and shows a 12 ton wagon built about 1922/23. It has side and end doors and Attock's 12 ton axleboxes. It is 16 ft 6 in. x 8 ft 0 in., with 5 x 7 in. and 2 x 9 in. planks and appears to be painted grey, with white letters and black shading. It is likely that this merchant supplied both the industrial and domestic markets, together with the fishing fleet. He would have obtained various grades of coal from Yorkshire/Derbyshire/Notts.

Plate 27: By the mid-1930s the 8 ton wagon was rapidly disappearing but the smaller merchants hung on as long as possible. Such an example, from an unknown builder, is shown here. It is fitted with side doors only, which, unusually for a five-plank wagon, do not extend to the top, and Ellis patent axleboxes. It appears to be 15 ft 0 in. x 7 ft 6 in., with 1 x 7 in., 1 x 5 in., 2 x 7 in. and 1 x 9 in. planks and is painted red, with white letters and black shading. The absence of the word 'merchant' from the livery probably indicates a plank repair. This trader could have obtained his supplies from around Wrexham or Flint, but more likely from North Staffordshire.

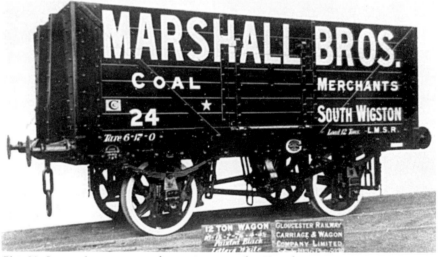

Plate 28: In complete contrast to the previous view this Plate shows a standard 12 ton wagon built in September 1935 and painted black, with unshaded white letters. The © mark to the left of the number shows that the Marshall Bros were party to the RCH commuted charge scheme. This entailed the payment of 1s. per wagon per annum in lieu of the shunting charge of 1s. per wagon and the siding rent charge of 6d. per wagon per day. However Messrs Marshall's were not listed in 1933 and may only have started in business with these wagons. It was more common for newcomers to buy second-hand 10 ton wagons cascaded down by the big companies modernising their fleets, but if one had the capital there was no law saying one could not start with spanking new wagons. South Wigston was on the MR Leicester-Rugby line and coal would have come from Leicestershire and South Derbyshire or Notts/Derbyshire.

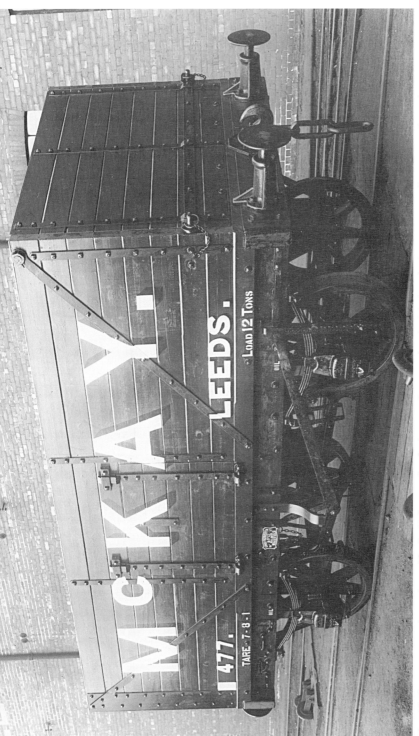

Plate 29: It was not usual practice at Charles Roberts to let shading overlap onto the plank below the lettering, as in this case. Perhaps the owner specified this, but the records for 1923-30 are lost so we will never know. The wagon is a standard 12 ton vehicle built in 1927, registered by the LMS and fitted with side, end and bottom doors. As such this trader would have supplied both the domestic and industrial markets in the Leeds area. The wagon was painted dark red, with white letters and black shading.

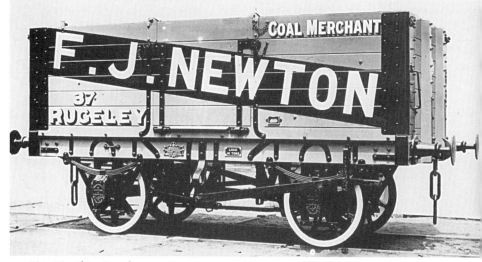

Plate 30: This view shows a 10 ton wagon with a very striking livery, built by Birmingham Carriage & Wagon Co. It has internal diagonal side braces, and is fitted with side doors only, with lift-over top door, brakes on either side and grease axleboxes. It also has buffer guides with foot-treads and commode handles at diagonally opposed corners. The wagon appears to be 15 ft 6 in. x 7 ft 5 in. with 1 x 9 in., 4 x 7 in. and 1 x 9 in. planks and is painted grey, with white letters and black shading. The main name has unshaded white letters on a black band. Rugeley lay on the LNWR Trent Valley line and supplies would have been obtained locally from Brereton Collieries and from North and South Staffordshire, South Derbyshire and Notts/Derbyshire.

Plate 31: Photographed shortly after overhaul in June 1930, this 10 ton wagon operated by a well known company in the London coal trade was built in 1902 and registered by the MR. It is fitted with side doors only, internal side braces and round bottom grease axleboxes. It appears to be 15 ft 6 in. x 7 ft 6 in. with 5 x 7 in. and 1 x 9 in. planks and is painted red, with white letters and black shading. The small lettering reads 'Postal address PEARSALL'S Ltd. 125, Westbourne Park Rd, W2'. While normally at home on the MR main line to the Notts/Derbyshire coalfield, the company's depot may well have been at Westbourne Park just outside Paddington.

Plate 32: The previous illustrations of London area wagons were large distributors, but of course there were merchants dealing exclusively with the domestic trade. An example is Nathaniel Pegg & Co., which had depots at Kensington (Warwick Rd) and Kew Bridge. Here much of his trade would have been in the Notts/Derbyshire Top Hard coal which was referred to as drawing room coal. The wagon shown was built in 1896, registered by the MR, and photographed just after overhaul by F.T. Wright, Nottingham, in August 1934. It is fitted with side doors only, united Ellis axleboxes, and probably as a result of collision damage it has had more modern ribbed buffer guides fitted to the right-hand end. It appears to be 15 ft 0 in. x 7 ft 6 in., with 5 x 7 in.. 1 x 5 in. and 1 x 7 in. planks, and is painted red, with white letters and black shading.

Plate 33: Illustrations of wagons indigenous to the West Country are quite rare, and this example also carries a maker's plate, the shape of which is quite unknown to the author. Photographed in the mid-1930s, the wagon was built around 1905-1910, and is fitted with side doors only with cupboard top doors, inside diagonal side braces and united Ellis axleboxes. When photographed it had a universal brake shoe (i.e. with fixing lug top and bottom, projecting up or down at 45°) at the left wheel and a handed brake shoe (i.e. right- or left-hand, fixing lug at top only, tapered bottom) at the right wheel. It appears to be 16 ft 0 in. x 7 ft 5 in., with 7 x 7 in. planks and is painted black, with unshaded white letters. Coal supplies to the Plymouth area would have come from South Wales, the Forest of Dean, Somerset, North Staffs and Notts/Derbyshire.

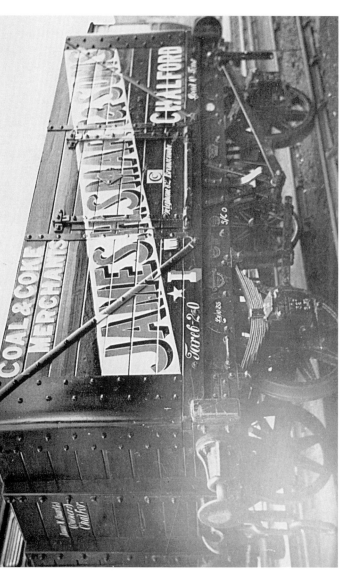

Plate 34: This highly colourful, but cluttered 10 ton wagon, owned by James H. Smart, Chalford, could be compared with Plate 30 as an example of how not to letter a wagon. Photographed in October 1935, after overhaul by Wagon Repairs Ltd at their Reparco Wagon Works, Lydney, the wagon is fitted with side doors only and Ellis patent axleboxes. It has elm packing washers to the buffer guides and side door springs, the smooth finish of which indicates they have just been fitted. The most striking constructional feature of the wagon, however, is the semi-elliptical (almost half-round) nature of the diagonal side brace, a feature never before seen by the author in studies of well over 1,500 photographs of PO wagons. Although difficult to read the register plate (Midland) indicates the wagon was built in 1906 and it was probably second-hand when acquired by Messrs Smart.

It appears to be 16 ft 0 in. x 7 ft 6 in., with 7 x 7 in. planks and is painted black. The main name is in red, with black shading on a broad yellow band. The smaller lettering is yellow, shaded red. Had the lettering been left at that it would have been nicely balanced, but there was the problem of the running number which was squeezed in on the bottom plank and the side rail. In small italics, probably in white, on the door is 'Telephone 82 Brimscombe'. Above this is the © mark, in black on a yellow rectangle. Finally to the left of the number is a yellow star, which indicates the wagon is subject to the provisions of the Commuted Empty Haulage Scheme.

In general terms PO wagons were not allowed to leave any station for any private wagon repairing shops without bearing the main line company's label showing the forwarding station. In addition a paid consignment label must also be obtained for such wagons, except those going to repairing shops in direct line of empty running to the colliery from which they were received loaded. Any wagon arriving at a junction with a 'foreign' line without labels showing fully the forwarding and receiving stations would be refused. If, however, they bore the yellow star, they would not be detained for the want of consignment notes or railway labels and would be forwarded immediately to the repair works. It is perhaps worth noting that these regulations did not apply to LNER lines in Scotland.

Plate 35: This view, taken at Rewley Road, Oxford (LMS) in the mid-1930s, shows the wagons of two old-established merchants. E. Welford & Son was in business in 1902, when Gloucester Carriage & Wagon Co. built them two 10 ton wagons, Nos. 38 & 40. The vehicle seen here, however is thought to be an ex-Caledonian Railway wagon. Built in the early years of the century, the wagon has side doors only, inside diagonal side braces, round bottom grease axleboxes and brakes on one side only, with a MR pattern long brake lever. It is 15 ft 0 in. x 7 ft 6 in., with 1 x 9 in., 2 x 7 in. and 2 x 9 in. planks and is red, with white letters and black shading. W. Simmonds & Son had been trading even longer, for the 8 ton wagon here, No. 25, was built by Gloucester in November 1897. It is fitted with internal side brakes, side doors only, No. 4 axleboxes, and again awaits its second set of brakes (made mandatory from November 1911). The wagon is 14 ft 11 in. x 7 ft 6 in., with 3 x 7 in. and 2 x 9 in. planks. Originally painted chocolate, with white letters and black shading, this may well have changed to the cheaper red in later years. Both these merchants would have obtained coal from Warwickshire and Notts/Derbyshire.

Plate 36: It is uncertain when Messrs Wood & Co. started trading, but it is known that by 1938 they had a fleet of 24 wagons. What seems fairly certain, however, is that they built up this fleet by acquiring second-hand wagons, as this motley collection on hire from Thomas Hunter, Rugby, suggests. Photographed in July 1929 after repainting, all three were built by Hunter, but they vary markedly suggesting three different former operators. The first wagon was built in 1910, registered by the LSWR, and fitted with side doors only, raised ends and Ellis patent axleboxes. It appears to be 16 ft 0 in. x 7 ft 6 in., with 5 x 8 in. planks. The most interesting feature of the wagon is its general state of health, for it has a pronounced longitudinal curve, which was known as the 'broken back syndrome'. Wagons with full height side doors did not have quite the same rigidity as those with through top planks and rough haulage or shunting, possibly coupled with some inferior timber or workmanship, could bring on the distortion. However providing the wagon was still basically sound it could run for years in this condition.

The second wagon, of similar age, has side doors only, with lift-over top door, internal side braces, Hunter's round bottom axleboxes and wooden side door stops. It has round based buffers but without the distinctive ribs seen on those on the first wagon. The lettering of the main name is in condensed style indicating a much shorter wagon and it is probably 15 ft x 7 ft 6 in. with 5 x 7 in. and 1 x 9 in. plank. The third wagon, No. 15, has side doors only, internal side braces, round bottom axleboxes and curved ends. It appears to be 16 ft 0 in. x 7 ft 6 in. with 1 x 9 in., 3 x 7 in. and 1 x 9 in. planks. All three wagons have brakes on either side and they are painted dark grey, with white letters and black shading. It is understood that this merchant was based on the former MR and his supplies would have come from Leicestershire, South Derbyshire, Notts/Derbyshire.

Plate 37: The mineral wagon was a work horse, not a thing of aesthetic charm, but this 10 ton vehicle, built by Hurst Nelson, around 1906-1910, is particularly neat and well balanced, both structurally and in terms of livery. It is fitted with side doors only, brakes on one side only and No. 33 axleboxes. It has a most unusual short side door stop, and a small striking plate on the second plank of the door. It is 16 ft 0 in. x 7 ft 6 in. with 6 x 7 in. planks and appears to be painted red, with white letters and black shading. This company was still in business in the late 1920s and would have purchased coal from Warwickshire, North Staffs and Notts/Derbyshire.

Plate 38: While they were in sound condition, and able to supply his requirements, the smaller trader hung on to the 8 and 10 ton wagons, as this view shows. Built around the turn of the century and registered by the MR, this 10 ton wagon, photographed in May 1938 after repainting, was fitted with side doors only, wooden doorstops and Ellis axleboxes. It was 15 ft 0 in. x 7 ft 6 in. with 1 x 9 in. and 4 x 7 in. planks(ends 2 x 8 in. and 3 x 9 in.) and was painted grey, with white letters and black shading. Coal would have been obtained from the Forest of Dean, Leicestershire, South Derbyshire or Notts/Derbyshire.

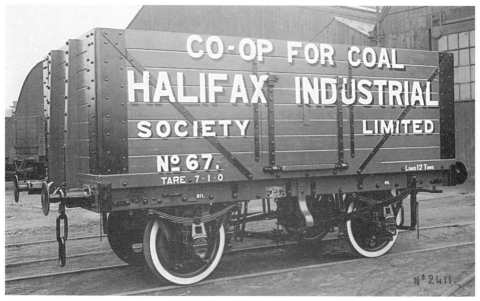

Plates 39 & 40: (*opposite*) The Co-operative movement began in Rochdale in 1844 to provide a means by which the ordinary man could obtain goods at the best possible prices. On payment of a small capital fee one could become a member of the Society, which purchased goods from the supplier and sold them to the members at market price. The profit made was then shared out to members, in the form of a 'dividend' in proportion to how much the member had spent. The ability to obtain goods at permanently discounted prices was naturally popular, and as coal was a relatively expensive item it was not surprising that Co-operative Societies entered the trade in a big way.

One such group was the Halifax Industrial Society which purchased its first wagon from Charles Roberts in April 1897. This was 15 ft 0 in. x 7 ft 6 in., with 5 x 7 in. and 1 x 9 in. planks, and was fitted with side doors only and No. 50 axleboxes. It was lettered as in *Plate 39*, numbered 36, and painted grey, with white letters and black shading. this was followed in March 1898 by a similar wagon, but with 7 x 7 in. planks and numbered 3. The years 1898 and 1899 saw the purchase of two identical wagons, Nos. 22 and 45, and this random selection of running numbers continued. In August 1900, four identical wagons, Nos. 11, 12, 40 & 41 were built, but the body colour was changed henceforth to Indian Red. These were followed in June 1901 by six wagons, Nos. 6, 7, 8, 13, 20 and 29 similar to the one of 1897, but with a wooden door stop either side. In August 1905 four wagons, Nos. 1, 4, 17, 21, were purchased. These were 16 ft 0 in. x 7 ft 6 in., with 7 x 7 in. planks, side and bottom doors only, and Attock's 46 axleboxes.

A further change was made four years later when Nos. 42, 43, 44, 52, 53 and 54 were built. They were dimesionally the same, but had 3 x 9 in., 1 x 8 in. and 2 x 7 in. planks, Rigley's axleboxes, side door springs and thresholds (the wooden strip on the side rail between the door hinges). The final batch of early wagons, Nos. 55-60, were ordered on 18th July, 1914, and were as the previous order, except for 7 x 7 in. planks, oil 126 axleboxes and brakes on either side.

With the outbreak of World War I in 1914 and the austere years which followed, the construction of new wagons just ticked over and it is interesting to note that in the period from 1897 to 1923, of the total output of wagons from Roberts, only some 6 per cent were built in the nine years after 1914. As far as Halifax Co-op was concerned it carried on with its existing fleet until 1929 when it acquired at least one new standard wagon, No. 46. On 11th April, 1931, three more wagons were ordered and once again changes were made. Apart from fitting of three-hole disc wheels the livery was changed to incorporate an advertising slogan 'Co-op For Coal' on one side, 'Coal At Co-op' on the other. The purchase of wagons from Roberts concluded, with two vehicles identical to No. 67 built in August 1935 (Nos. 61 & 63), and No. 32 built in September 1936.

Bearing in mind the number of coal merchants in Halifax and the Colne Valley, and allowing for the replacement of early wagons, the size of the Halifax Co-op fleet was inordinately large for the size of the domestic market, and the purchase of bottom door wagons from 1905 to 1914 suggests they also served the industrial market. The local mill owners, being frugal types, would be only too keen to join the Co-op, get their dividend and increase their profitability, thus these wagons would have visited different collieries in South and West Yorkshire for various classes of coal.

Plate 41: The fact that this wagon built in June 1904, bears the number 1 suggests that this branch of the Co-operative movement had just entered the coal trade. The wagon is a typical Gloucester product with side doors only and No. 48 axleboxes, but interestingly it only carries one of the well known big G solebar plates, which tells us that the wagon had been purchased outright by the Co-op. Although slightly out of focus the register plate appears to read 'NSR', and one would certainly expect the majority of coal to have come from North Staffordshire. However the wagons would also have travelled via Rowsley and Toton to Notts/Derbyshire Collieries.

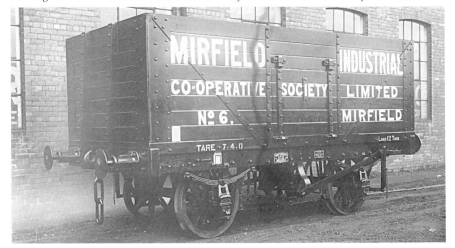

Plate 42: Mirfield Co-op had its first wagons Nos. 1, 2, 3 from Roberts in 1896 and 1900. These were dark red and lettered 'PERSEVERANCE CO-OPERATIVE SOCIETY MIRFIELD'. They were replaced in August 1902, with three reconstructed wagons, also Nos. 1, 2, 3, and painted red, lettered as the wagon illustrated. The story then goes stone cold until No. 6 popped up in 1928. It was painted black, with unshaded white letters and ran to collieries in South and West Yorkshire.

Plate 43: (*opposite*) For every ton of coal carbonised to make domestic gas, 14 cwts of coke was produced, and this was sold in the open market. In the early years of the century, when coal was around 10 to 11 shillings per ton, coke was selling at up to 13 shillings per ton, and this may have persuaded Morley Corporation to push these sales by the acquisition of a coke wagon. This would enable them to deliver to a wide range of customers, including those without wagons, throughout West Yorkshire and beyond. Ordered on 20th May, 1908, the wagon was fitted with side doors only, brakes on one side only, Attock's 46 axleboxes and 1 ft 10 in. coke grating. It was 15 ft 0 in. x 7 ft 6 in. with 7 x 7 in. planks and was painted grey, with white letters and black shading.

Chapter Four

Public Utilities and Industry

In 1938 some 27,000 coal wagons were operated by public utilities, the coking and metallurgical trades, general industry, food manufacturers and others outside the mining and retailing of coal. One of the largest constituents of this group was the gas and electricity undertakings, of which there were well over one thousand, a considerable proportion being owned by local authorities. Many of these works were situated away from the railways, necessitating road haulage of raw materials in, and by-products out, to the nearest goods yard. Nevertheless about 10 per cent of the undertakings, split almost equally between private and local authority control, operated their own wagons. Irrespective of wagon ownership the utilities as a whole consumed about 22 million tons of coal annually, and, in addition to gas and electricity, produced solid and liquid chemical by-products and around 15 million tons of coke suitable for domestic use. Some of this was sold locally, particularly in the largest cities, but a considerable proportion was purchased by contractors and merchants and taken away by rail.

With such a large number of undertakings under local authority control, the amount of detailed information available in the form of Minutes is staggering and way beyond the scope of the present narrative. Nevertheless the compilation of this manuscript has prompted the author down that avenue of research and a tantalising glimpse is given at *Plates 44 & 45*.

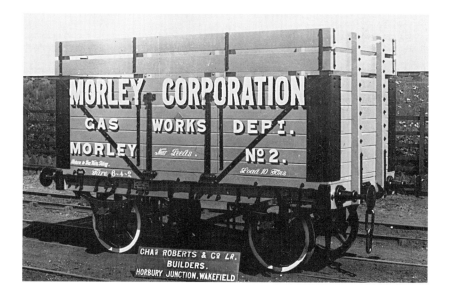

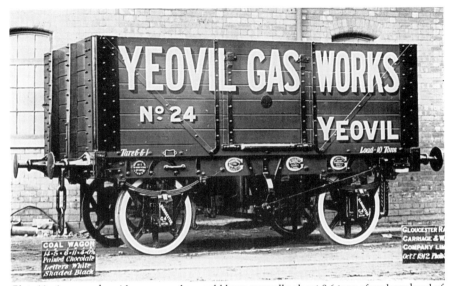

Plate 44: As a rough guide a gas works would burn annually about 0.6 tons of coal per head of population in its area of distribution, which would give an annual requirement at Yeovil, in the early years of the century, of approximately 1,000 tons. The normal method by which such fuel supplies were obtained consisted of annual contracts with distributors, who would handle far greater quantities of coal than their own wagon fleets could cope with. In what was a highly competitive market place it was imperative that the supplier put in the best possible price. If he could avoid the need to hire colliery or railway company wagons, by using the customer's own wagons, he could reduce prices by 4-5 per cent. On the other side of the coin, a customer with fixed, long term requirements for coal, would be anxious to retain lower prices, particularly if this could be achieved by short term investment. In 1905-10 the purchase price of a new wagon was £90-£100 and a 4½ per cent saving on the cost of coal would pay for this over 4-6 years, dependent on the fluctuating price of coal. Thereafter, subject to maintenance and repair costs, this reduced price of coal could be sustained. Even if an operator did not wish to purchase wagons outright he could still achieve a saving of about 2½ per cent on the cost of coal by taking wagons on simple hire from the builders, inclusive of repairs, at much more favourable rates than hiring from the colliery or railway company.

Contrary to public opinion many local authorities were, and still are, efficiently run and managed and Yeovil Council was an early advocate of wagon ownership. The total size of its fleet has not yet been ascertained but on 1st September, 1903 the Council ordered six wagons from Charles Roberts. These had side doors only, Attock's 46 axleboxes and measured 15 ft 0 in. x 7 ft 6 in., with 7 x 7 in. planks. Numbered 15-20 and registered by the GWR, they were painted chocolate and were lettered:

YEOVIL CORPORATION
GAS   DEPT

No   YEOVIL

Above 'YEOVIL', in italics was *'For Repairs advise Gas Manager'*. These wagons were delivered new to Newton Chambers & Co. to load gas coal.

The wagon illustrated, identical to No. 8, (tare 6-2-1) built in December 1910, was built by Gloucester Carriage & Wagon Co. in October 1912 and was fitted with side doors only, brakes on either side and No. 4 axleboxes. The buffers had foot treads and diagonally opposed commode handles were fitted on the third and fourth planks of the ends. They were painted chocolate, with white letters and black shading and it is likely that the revised style of lettering was applied to the earlier Roberts' wagons as they came up for periodic overhaul about this time. It is interesting to note that the Gloucester-built wagons were registered by the LSWR. They were also on simple hire as indicated by the three plates on the solebar. The left one is the repairs' plate, the centre one the owner's plate and the third one the builder's plate.

When the wagon illustrated was built coal supply was largely in the hands of Renwick, Wilton, Torquay; John Brown & Co., Rotherham (through Walter Moore & Sons, London) and Newton Chambers & Co., Sheffield. By 1930, which was chosen at random, records had become more sophisticated and it was resolved to accept the following tenders:

G. Bryer Ash, Portsmouth 500 tons Denaby Nuts	@	29s.	7d.	per ton
Evesons Ltd, Birmingham 2,000 tons Wath Main Nuts	@	30s.	6d.	
Thos. Cash & Co. Ltd, Birmingham 2,000 tons Askern Main Nuts	@	29s.	11d.	
J. Longbottom & Sons Ltd, Sheffield 2,000 tons Manvers Main Nuts	@	30s.		
E. Foster & Co. Ltd, London 500 tons Elsecar Nuts	@	29s.	4d.	

(This traffic would have flowed via Annesley, Woodford Halse, Banbury, Reading, and Basingstoke)

Renwick, Wilton & Co., Torquay 750 tons Clay Cross gas coal	@	28s.	6d.
Lowell Baldwin, Bristol 2,500 tons Silverhill Nuts	@	29s.	4d.

(This traffic would have gone via Toton, Washwood Heath, Westerleigh (Bristol), Bath & Templecombe)

The average price of 29s. 8d. per ton for this coal admirably illustrates the effect of carriage charges, for it is almost 50 per cent higher than the cost of the same coal, at the same time, in Derbyshire.

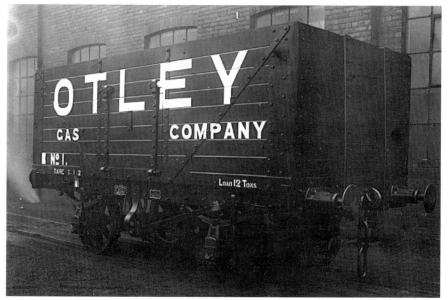

Plate 45: By no means have all company records survived, and little is known about Otley Gas Company, other than that it was in a group which also contained Harrogate and York Gas Companies. It is likely that the company was modernising its wagon fleet when it acquired this standard 12 ton wagon from Charles Roberts in 1928. It is painted black, with unshaded white letters, and is a good example of a wagon where stencils have been used to apply the Tare and Load details. In the 1930s the works would have carbonised around 6,000 tons of coal per annum and this would have been acquired from collieries in South and West Yorkshire and north Derbyshire. With several gas works in one group there would have been a certain amount of wagon pooling and it is conceivable that on occasion the wagon illustrated travelled to both Harrogate and York.

Plate 46: The juxtaposition of limestone, sand and coal in the lower Aire valley was a prime reason for the development of glass bottle manufacture in the Castleford-Knottingley area of Yorkshire. Two of the companies involved had their own wagon fleet, built in part, at least, by Charles Roberts.

The first wagons for Bagley & Co. comprised a batch of 12, numbered 71-82, ordered on 31st January, 1905. These were 16 ft 0 in. x 7 ft 6 in., with 7 x 7 in. planks and were fitted with side and bottom doors, brakes on one side only and Attock's 46 axleboxes. They were painted red, with white letters and black shading and the livery was very similar to that illustrated. On the top three planks was 'BAGLEY' (20½ in.), and the running number and 'KNOTTINGLEY' (7 in.) was on the bottom plank. The remaining lettering (6½ in.) was spread evenly over the remaining three planks. A second batch of ten 12 ton wagons, Nos. 20-29, were built in April 1916. These had side, end and bottom doors, brakes on either side, side door springs and Attock's 158a axleboxes. They were 16 ft 6 in. x 7 ft 11 in. with 5 x 7 in., 1 x 8 in. and 1 x 9 in. planks and were painted as the previous wagons.

While some of the early wagons may have been replaced in the 1920s this cannot be confirmed, but the wagon illustrated was built as a one-off in June 1931. It was fitted with side and end doors, and three-hole disc wheels. It was painted red, with white letters and black shading.

Plate 47: (*Previous page*) Messrs Jackson Bros' Headland Glass Works was situated a little to the south-west of Bagley's, on the L&Y & GN joint Doncaster-Pontefract railway. The earliest wagons built by Roberts were Nos. 11-14, erected in February 1911. They were 16 ft 0 in. x 7 ft 6 in., with 7 x 7 in. planks, were fitted with side doors only, with cupboard style top doors, brakes on one side only, side door springs and Attock's 46a axleboxes. They were painted red, with white letters and black shading, with 'JACKSON BROTHERS' on the top three planks, and the remaining lettering one plank lower than on the wagon illustrated.

By 1915 the company had changed its status to Jackson Brothers (of Knottingley) Ltd. In February of that year four more wagons, Nos. 15-18, were hired from Roberts. These were almost identical to the earlier ones, but had brakes on either side and Attock's 9a axleboxes. The wagons were delivered new to Bullcroft Main colliery and would normally have travelled between the works and collieries in the Doncaster, Wakefield and Barnsley areas.

Plate 48: Before World War II almost every town of any note produced its own beer, but Burton-on-Trent was the unchallenged centre with no less than 16 breweries. As the 20th century advanced more and more brewers took to malting their own barley, to suit their particular brews. However, following instances of arsenic poisoning in the early 1900s it became customary to use only anthracite, which contains less than four parts per million of arsenic trioxide, to generate the heat for this process. Of the remaining stages in brewing, mashing and boiling both used heat, but as contact with the product was more remote, good quality steam coal was adequate.

To transport these supplies many brewers operated their own wagon fleets, the wagon illustrated being from that serving Ind Coope & Allsopp Ltd. Relegated to internal use only when photographed in the 1950s, the wagon was built about 1905-1910 by Thomas Burnett, Doncaster, and as seen is fitted with side doors only, brakes on one side only and round bottom grease axleboxes. Somewhere along the line, the right-hand axlebox, on the side photographed, has been replaced by a united Ellis box, while side door springs and ribbed buffers have been fitted. The majority of ironwork nuts are on the outside, the usual practice, but those on the side door washer plates, and the abnormally large door striking plate, are on the inside. The right-hand corner plate carries the usual double row of nuts and has the diagonal side brace on the outside, as one would expect. The left-hand corner plate, however, has one staggered row of nuts and overlays the diagonal, strongly suggesting that the wagon has been converted from an end door type. It measures 15 ft 0 in. x 7 ft 6 in., with 5 x 7 in. and 1 x 9 in. planks and is painted red, with white letters and black shading. The wagon would have travelled to South Wales for anthracite, or to Leicestershire and South Derbyshire, Notts/Derbyshire or North Staffordshire for steam coal.

Plate 49: Not all private owner wagons carried ornate liveries as this example shows. While the actual maker is not known, it was built in 1902, registered by the Caledonian Railway, and fitted with cupboard style side doors, bottom doors, end doors with D-dropper fastening and round bottom axleboxes. Hanging from the bottom of the left-hand door can be seen the side door fastening wedge. To the right of this hangs the monkey-tail peg and chain. When this was withdrawn from its bracket, the monkey tail, just visible behind the door wedge, could be pushed inwards to open the bottom doors. A further intriguing feature of the wagon is something like five or six washers behind the nuts on the strapbolts (protruding through the headstocks) and buffer securing bolts.

The wagon appears to be 16 ft 0 in. x 7 ft 6 in., with 4 x 9 in. and 1 x 11 in. planks and is probably painted red. Both coal and limestone were used in alkali manufacture, but the five doors to this wagon suggest it was a coal wagon. This being so it would have travelled from Winnington to collieries in Notts/Derby or North Staffordshire.

Plate 50: The Staveley Coal & Iron Co. was established in the 1860s. By 1930 the company had nine collieries, two blast furnaces, a set of coke ovens and various chemical works. In addition to pig iron and coke, the company produced a variety of iron castings, Sulphate of ammonia (ammonium sulphate), Muriate of ammonia (ammonium chloride), pitch, Benzole (benzene), Creosote, and Carbolic acid (phenol). It also produced such household items as Caustic soda (sodium hydroxide) and bleaching powder, which were advertised as part of its wagon liveries. Seen here on 4th October, 1933 this 12 ton wagon is fitted with side and end doors, self contained buffers and united Ellis axleboxes. Built 10 years earlier by the Staveley Co. it was 16 ft 0 in. x 7 ft 11 in., with 1 x 9 in. and 6 x 7 in. planks. It was painted grey, with white letters and red shading. The running number, tare (7-3-1) and cross symbol were white on black.

Plate 51: From the same photograph as the preceding plate, this 12 ton wagon was built by Hall Lewis in June 1924 and was fitted with side and end doors and united Ellis axleboxes. It was 16 ft 6 in. x 7 ft 10 in., with 6 x 7 in. and 1 x 9 in. planks and was painted black, with white letters and red shading. While these wagons, which were typical of the Staveley fleet, would have carried coal to the company's works, they would also have been in general use. As a group the collieries produced house, steam, manufacturing, gas and coking coal, which enjoyed a wide inland market and was shipped at Hull, Grimsby and Immingham.

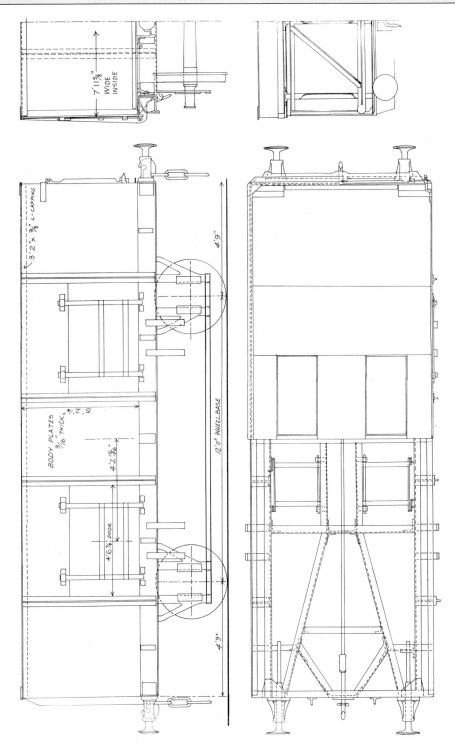

General Arrangement 20/21 Ton All-Steel Wagon

Chapter Five

Larger Capacity Wagons

Having looked broadly at the transport of coal, and before proceeding to other topics, it is perhaps opportune to examine the case put forward for the introduction of larger wagons.

In the early 1920s there were roughly 116,000 private owners' coal wagons in South Wales, primarily of 10 ton capacity, and the congestion caused by traffic to and from the docks to the collieries was becoming intolerable. The question of wagon capacity had been discussed along with that of wagon ownership and operation for many years but in September 1923 Sir Felix Pole, the GWR General Manager, addressed the problem in earnest, and wrote to the South Wales coal and mineral traders in the following manner:

Under the Railways Act, 1921, the railway companies and the traders are directly and mutually interested in efficient working of the railways. In considering ways and means of operating the traffic more economically, an outstanding point for consideration is that of increasing the carrying capacity of wagon stock.

For some years past the Great Western Railway have experimented with 20- and 30-ton wagons for general merchandise, but there are many difficulties in a country like Great Britain in using high-capacity wagons for this traffic, which passes in relatively small consignments.

As regards coal and mineral traffic - which constitutes 70 per cent. of the total tonnage conveyed by them - the Great Western Railway Company are satisfied that there should be no further delay in introducing wagons of at least 20 tons carrying capacity. The advantages of a 4-wheeled vehicle capable of carrying 20 tons of coal have been demonstrated from experience in using wagons of this capacity for the conveyance of locomotive coal.

The cost of constructing two 10-ton ordinary coal wagons is roughly 50 per cent. more than the cost of building a 20-ton wagon, and, as the census taken by the Board of Trade in 1918 disclosed that there were roughly 116,000 private owners' coal wagons in South Wales (mostly of 10 tons capacity), it is obvious that the introduction of 20-ton wagons would effect a very large saving in the annual costs of renewals. In addition, the necessity for more siding accommodation at the collieries and on the railway would be largely obviated, while the cost of repairs to wagons and sidings would be reduced.

In the interests of the trade of South Wales - where the practice in this respect is behind that on the North-East Coast - the owners of the coal wagons are invited to commence the construction of 20-ton wagons as soon as possible. It is recognised that before any change can be made many difficulties have to be overcome at the collieries and at the docks, but the present is an opportune time to consider the proposal, in view of the fact that a number of traders are understood to be placing orders for the renewal of their rolling-stock, while arrears of maintenance of tipping appliances &c., at the docks are being taken in hand.

Irrespective of the initial saving in cost of wagon construction and the fact that railway operating economies will not accrue until a considerable number of high-capacity wagons are available, the Great Western Railway are prepared AT ONCE to allow a rebate of 5 per cent. off their rates in respect of coal class traffic conveyed wholly over their system in fully-loaded 20-ton wagons.

Will you be good enough to let me know whether you are prepared to go into the matter with the company, in which case I will arrange for representative officers to wait upon you to discuss details.

Later, in 1924, in a further attempt to encourage the construction of 20 ton wagons, the GWR offered a reduction from 7½d. to 6d. per ton in shipping and weighing charges at the docks on coal shipped from 20 ton wagons. Simultaneously the company instigated a scheme for the provision of 20 ton wagons for hire, and embarked upon a programme of adapting shipping appliances at the docks to take these larger wagons. Despite these efforts the response of the coal owners could be succinctly described as disappointing.

Some years later the argument was taken up by vested interests, in the shape of the British Steelwork Association, which published a memorandum on Modern Mineral Transportation. It opened by stating that while the last 30 years had seen profound and far-reaching changes in the organisation of industrial development, the transport of minerals had remained prodigally wasteful of time, space and money. To maintain its status as a major manufacturing force the country's works , factories, gas works and power stations needed cheap supplies of power in the form of coal, yet low pit-head prices were pointless if excessive rail transport costs applied. The export trade in coal, competing on the world market, and the inland domestic trade, competing with imported oil, also needed efficient cheap transport.

These arguments could not be disputed, nor could the simple truth that far too high a proportion of the mineral wagon stock was of great age, and to some degree of obsolete design, with a small unit capacity and disproportionately high tare weight. it was also true to say that the basic design of the wooden mineral wagon necessitated haulage at low speed, and that the condition of far too many wagons was such that even slow schedules were difficult to maintain due to stoppages for repairs. The low unit capacity not only increased the tare weight (or non paying load) which had to be hauled, but increased shunting movements and produced congestion at sidings.

If conditions were to be improved a large proportion of the wagons in use, many of 10 ton, or even 8 ton capacity, would need to be withdrawn and replaced by modern stock of at least 20 ton capacity. It was clear that this must be done on a large scale, for the inclusion of but a few old 10 ton wagons in a train of 20 ton vehicles would require that the train be slowed down to suit the oldest, and weakest, vehicles. The Association did accept, however, that for certain services the 12 ton wagon must be retained, but qualified this by stating that they should be of a modern design.

A large majority of mineral wagons were in the private ownership of collieries, factors, merchants, manufacturers etc., and it would appear at first sight that the then-current state of affairs was caused by the private owner's failure to accept that his stock was obsolete. The crux of the matter was that the owner who purchased high capacity wagons and loaded, say, 1,000 tons of coal in 50 wagons (weighing 500 tons), would still be charged the same rate per ton as another who loaded 1,000 tons in 100 smaller wagons weighing 650 tons, which required more motive power and occupied greater running and siding space. The economies achieved by hauling single trains of high capacity would

accrue solely to the railway companies, the benefits to the wagon owner of
reduced maintenance and repair costs on new wagons being inadequate on
their own to justify the capital expenditure on such wagons.

It has already been noted that the GWR had offered a 5 per cent rebate on
freight charges for the use of 20 ton wagons, but this had not been a significant
success. With this situation, and bearing in mind that for some purposes the 12
ton wagon was most appropriate, the Steelwork Association proposed that the
rebate should be set at 5 per cent for minerals carried in 12 ton wagons and 10
per cent for the use of fully loaded 20 ton wagons. Here the vested interest
surfaced again for it was claimed that such a move would not only free the
railways from the burden of a large number of obsolete wagons causing delay
and congestion, but the construction of new wagons would be a powerful
stimulus to productive employment. Naturally, if the construction of new
wagons was to be encouraged it was essential that the most modern methods
and materials should be used and this would necessitate all-metal construction,
for the following reasons:

1. All-metal wagons were greatly superior to wooden or composite (wooden body on
   steel underframe) construction.
2. The building of all-metal wagons afforded more employment to British labour,
   stimulated British industry, and affected favourably the balance of trade by
   reducing imports of foreign timber.

The superiority of all-steel wagons was generally accepted. The sturdier
frame could better withstand heavy shunting, reducing maintenance costs and
prolonging the life of the wagon. It was held in certain quarters that the
carriage of freshly washed coal would cause excessive corrosion, but provided
the wagons were repainted at regular intervals (as wooden wagons were), this
claim could not be substantiated.

Apart from the operational advantages claimed for the use of 20 ton wagons,
there was a wider economic argument for the building of all-steel wagons.
Instead of depending on imported timber, the iron and steel of all metal wagons
was entirely produced by British labour from British raw materials, thereby
increasing employment. For every ton of British steel produced, no less than
seven tons of raw materials - iron ore, coal, coke, limestone etc. - are consumed,
and the transport of these materials directly benefited the railway companies.

Objections were raised that while the mass construction of all-metal wagons
would stimulate employment generally, it would suppress employment
directly involved in the construction and repair of wooden-bodied wagons.
However, four-fifths of the labour employed in wagon building shops would be
unaffected, and the greater part of the remainder, particularly the younger men,
could be retrained in the new methods of construction. This question of
industrial stimulation and employment was of paramount importance, for the
whole argument by the British Steelwork Association was being put forward in
the early 1930s when government and industry were desperately seeking ways
to break the depression and mass unemployment. The government played its
part by giving financial assistance to the GWR under the Development (Loan

Guarantees and Grants Act) 1929, in its attempt to encourage the use of 20 ton wagons, and it was only just that other railway companies and private owners could expect similar assistance. It would also be necessary for the railway companies to play their part by offering more attractive rates for the carriage of minerals in newer, larger capacity wagons.

Despite all the foregoing arguments there were some cold statistics to be faced. While only 7 per cent of railway company stations and coal depots could not handle 20 ton wagons, 64 per cent of collieries, 55 per cent of dockside loading appliances, and 45 per cent of private sidings serving works with a weekly consumption of coal of 500 tons or more, could not take such wagons. In fact, after several years persuasion, a census on 31st January, 1928 showed that out of a total of 578,626 open coal and coke wagons, a mere 1,709, or 0.3 per cent, had a capacity of 20 tons or more. During the early 1930s wagon builders continued to stimulate demand by hiring out sample wagons, but the 'bottom line' was that colliery owners, factors, merchants, manufacturers and public utility undertakings were happy with their traditional wagon fleets and would not, or could not, embark on their wholesale replacement by 20 ton all-metal wagons. Such action was only to follow many years later under a state owned railway system.

Nevertheless, 20 ton all-metal wagons were part of the private owner scene and it would be remiss not to illustrate these.

Plates 52 & 53: (*Above and opposite*) On 4th March, 1931 management at Charles Roberts placed an order in the book for five 20 ton all-steel wagons, and it is understood that each one was painted for a different operator and hired out for evaluation trials. The 20 ton wagon was built with various door arrangements, but this batch had two side doors, four bottom and two end doors, suggesting that they were intended primarily for use in the shipping trade. The wagons were 21 ft 5¾ in. x 8 ft 0 in. x 5 ft 2 in., with side stanchions 4 ft 1½ in. and 8 ft 3 in. from the ends, and with a 12 ft 0 in. wheelbase. One wagon (*above*) was painted bright red and lettered in white with black shading for Bullcroft Colliery. A second wagon was painted black and lettered for South Kirkby, Featherstone and Hemsworth Collieries Ltd., The small lettering reads 'Empty to South Kirkby Colliery Sidings LNE LMS', and this was probably the only colliery in the group where the screens could accommodate such wagons.

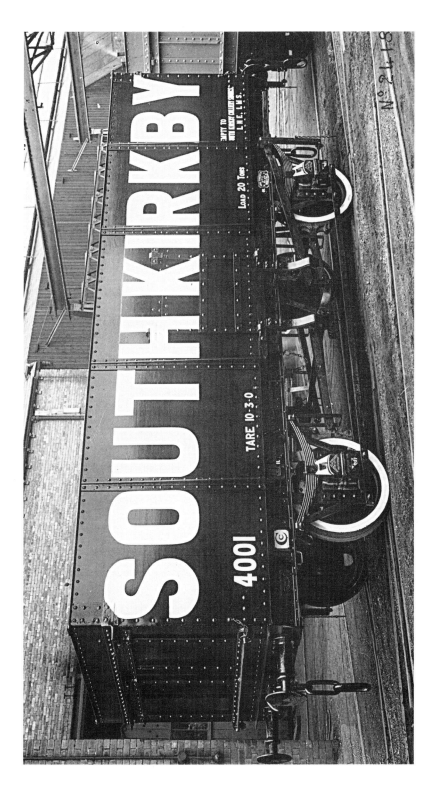

SOUTHKIRKBY

4001

TARE 10·3·0.

LOAD 20 Tons

EMPTY TO
WILL KIRKBY COLLIERY SIDINGS
L.N.E. I.M.S.

Plate 54: As the 1930s progressed the all-steel wagon became more widespread and in 1939 Charles Roberts built 1,515 such wagons, of various tonnages, including fifteen 21 ton vehicles for Messr Charrington, Gardner and Locket, ordered on 21st May and numbered, it is believed, 7300-7314. These were to the same overall dimension as shown in *Plates 52 & 53*, but were fitted with four side doors only and standard oil axleboxes, with the initials CGL cast on. They were painted red, with white letters, with name and, number shaded black. The small lettering reads 'Charrington, Gardner, Locket & Co. Ltd. 16 Mark Lane E.C.3.' A wooden panel affixed at the left-hand end of the body carried separate tin plates bearing the Ⓒ mark, a yellow star on a black base (indicating that the owner participated in the 1933 commuted charge scheme for the haulage of empty wagons) and a loading label. In this case, this reads 'Empty to ORMONDE Colliery LMSR to load for LPTB private sidings Cremorne LMS GWR.' (Cremorne Wharf was at Chelsea Basin.)

Plate 54a: Detail of V-hangers and link mechanism fitted on one side of this type of wagon.

Chapter Six

Other Mineral Wagons

While the great majority of private owner wagons were used for the carriage of coal, 8 per cent, or around 50,000, were used for other purposes. These included a handful of covered vans for salt, lime and manufactured goods; tank wagons for fuel, lubricants and chemicals and open wagons for miscellaneous minerals. Crushed limestone from Derbyshire, Yorkshire, Leicestershire, Somerset; the famous Portland and Bath building stone, Derbyshire gritstone, and granite from Cumberland and North Wales were carried everywhere. Slate, sand, gravel, china clay and iron ore also needed transport. Semi-manufactured or waste goods such as macadam, tarmacadam and slag all required wagons, as did bricks and refactory goods. Some variety!

Plate 55: Although the Staveley Company had its own wagon building shops it did not build this particular vehicle, which must have been hired for the transport of materials used in extensions to the brickworks in the early 1930s. It is fitted with side doors only, Oil 116 axleboxes, hockey-stick knee washer plates and round base ribbed buffers with foottreads.

It is 16 ft 6 in. x 7 ft 11 in., with 2 x 9 in. and 1 x 7 in. planks and has a tare weight (below the number) of 5-7-4. It appears to be painted dark grey with unshaded white letters.

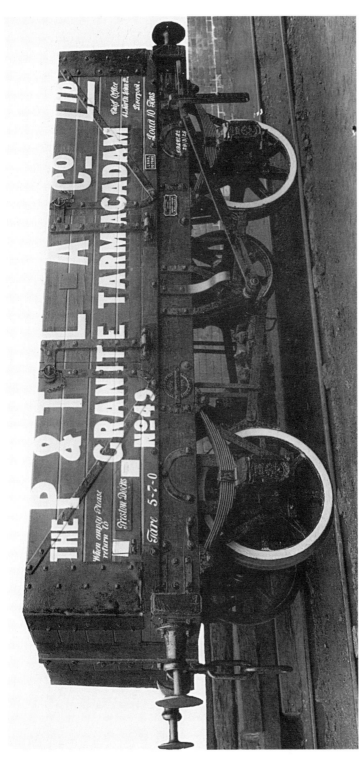

Plate 56: One of the popular arguments put forward for the adoption of modern all-steel wagons was that there was a general air of obsolescence about the older wagons. Old this wagon may be, having been built in 1892 and registered by the MR (registration no. 11074), but dilapidated it is certainly not. Photographed after a full overhaul was completed at Roberts on 29th March, 1926, the wagon is fitted with side doors only, brakes on either side and Attock's 46a axleboxes. The paint finish on the door stop and springs suggest that these have just been fitted and renewed respectively. Three separate quarries operated at Penmaenmawr, North Wales from around 1830 and these merged to form the Penmaenmawr and Welsh Granite Co. in 1911. Much of the output of the quarries was shipped to Bristol, London, Manchester and Preston. In 1921 the company amalgamated with the Limmer Co. to form the Penmaenmawr & Trinidad Lake Asphalt Co. Carrying road building materials, the wagons would have travelled widely in northern England, Wales, the West Midlands etc.

The wagon was 15 ft 0 in. x 7 ft 6 in., with 3 x 9 in. and 1 x 7 in. planks and was painted red, with white letters and black shading. The small italics at the right end read *'Chief Office, 41 North John St, Liverpool.'*

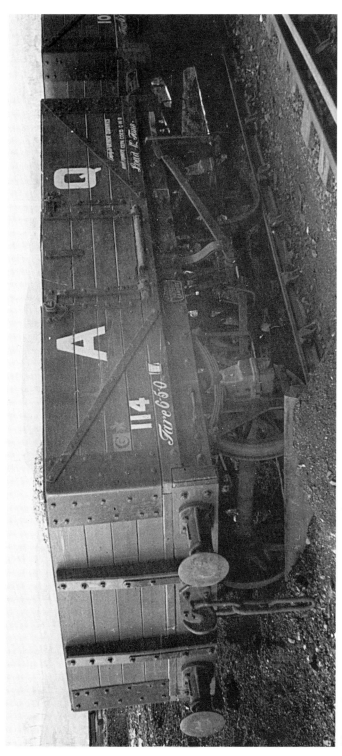

Plate 57: In some cases it is difficult to ascertain the ownership of a wagon from its livery, and while this had been ascertained here, the meaning of the letters 'A Q' remain a mystery. The wagon was originally owned by Crawshay Bros, whose interests included Vaynor Limestone Quarries, ironworks and mining around Cefn Coed in Glamorgan. By the late 1920s the company had become part of Llewellyn (Cyfarthfa) Ltd, of Merthyr Tydfil, which in turn became part of the giant Gueret, Llewellyn & Merritt, of Cardiff. It is not known, unfortunately, if the wagons ever received the well known GLM insignia.

The wagon is built to the 1923 RCH specification and is fitted with side doors only, standard running gear and buffer cylinders with foottreads. It is 16 ft 6 in. x 8 ft with 3 x 7 in. and 1 x 9 in. planks and is painted red, with white letters and black shading. The small lettering reads *'Empty to* VAYNOR QUARRIES *and repairs* CEFN COED GWR.'

Plate 58: At the end of the 19th century the Peak District limestone industry was in dire straits, with a large number of companies in direct competition, over-production, and low prices, which combined to depress profits and prevent development and expansion. Thus when H.A. Hubberty, of the Buxton Lime Co., suggested an amalgamation he found sympathetic ears and in 1891 thirteen quarry owners came together to form Buxton Lime Firms Co. Ltd. At the time many of the quarries are believed to have had their own wagon shops and thus a hotch-potch of wagons came into the new group. While BLF did commence to build wagons to its own design, the inherited fleet was of remarkable longevity. One such wagon, almost certainly converted from dumb buffers, and seen here in 1919, is fitted with side doors only and self contained buffers. It would have had grease axleboxes and may well still have had brakes on one side only. It also has very unusual corner plates in two sections. More importantly, however, the wagon has two features, which to the author are unique. It has inside diagonal side braces, common enough, but in this case they run from the top of the side-knee to the bottom of the corner plate. Secondly, while the left-hand door hinge has a conventional fastener of a wedge through a hasp, the right-hand one has a small clasp which appears to rotate about a central bolt. The wagon was probably 14 ft 6 in. x 7 ft 5 in., with 3 x 7 in. and 1 x 9 in. planks at the side and 5 x 7 in. and 1 x 9 in. at the ends. As seen, the paintwork on the body, has faded completely and the white paint of the lettering has virtually gone to leave the letters 'BLF' in the original grey of the bodywork. While the number has been repainted the rest of the wagon has not seen a paintbrush since the shortened livery was applied around 1901/02.

Plate 59: Limestone has many uses, perhaps today dominated by aggregates and construction, but 70 or 80 years ago the picture was different. It was in widespread use in wire-drawing, water purification, glass making, iron and steel making, agriculture etc. With respect to the latter, the sugar beet factory at Kidderminster, one of 18 operated by the British Sugar Corporation in 1939, alone used 8,000 tons of Buxton lime during the sugar beet season. Such tonnages, however, were insignificant when compared with the demand from the rapidly expanding chemical industries of Cheshire and elsewhere. Such was the dependence of the industry on limestone, that in 1919 Brunner Mond, one of BLFs most important customers, acquired a controlling interest in the company, which in turn became part of ICI on 1st January, 1927.

Buxton Lime Firms never invested in a modern wagon fleet but kept its wagons in good repair, and in its early years ICI carried on this tradition. Wagon No. 48 dates from the early 1890s and is seen here after overhaul at the Harpur Hill workshops in November 1934. It is fitted with side doors only, brakes on either side, square bottom grease axleboxes and internal side braces, which are bent at the bottom to avoid the axleguard bolts. The wagon was 14 ft 11 in. x 7 ft x 5 in., with 3 x 9 in. planks on the sides and 3 x 9 in. and 1 x 11 in. planks at the ends and was painted grey, with white letters and black shading.

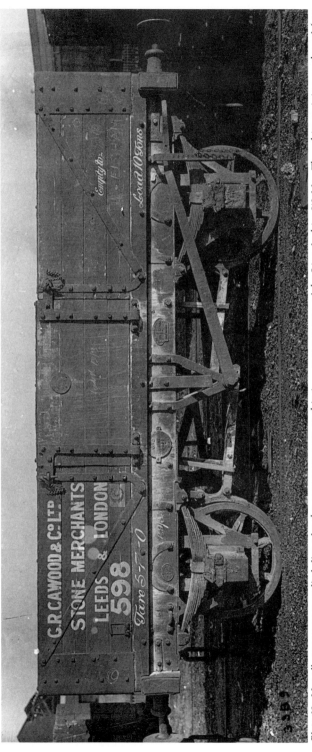

Plate 60: Not all stone was supplied direct by the quarry owners, as this 4-plank 10 ton wagon operated by G.R. Cawood & Co. Ltd shows. This is yet another example of longevity, for the wagon was built in 1892 and registered by the L&YR. It is seen here about November 1939 at Roberts, where the chalk crosses on the left-hand headstock, and the obvious sag of the associated buffer ram suggest it has come in for attention following a heavy shunt. The oval plate above the left-hand axlebox shows the vehicle had had a general repair in December 1937, but it was not repainted, for the paint date 9/35 is just visible to the right of this. On the original print, below the words *Empty to*, can be seen chalked 'ELSHAM', which is between Thorne and Barnetby, close to both the chalk and oolitic limestone escarpments of the Lincolnshire Wolds. The white staining on the solebar and running gear would have been caused by the adherence of dust being washed down between the floor planks and the siderail.

The wagon is fitted with side doors only, brakes on either side, plain buffer guides and Ellis axleboxes, the right-hand visible one being a Roberts' product. It is 15 ft 0 in. x 7 ft 6 in, with 4 x 7 in. planks and appears to be painted black, with plain white letters. When photographed the wagon was due for its periodic repainting, but World War II had broken out. It is impossible to say whether the habits of turning out pristine wagons had not yet been nullified by battle, or if the urgency of returning wagons to service had already filtered down to Horbury Junction.

Plate 61: Before brick, concrete and glass became standard building materials, anything more than a small factory or house, and many civil engineering features, were built in stone. This was produced in many sizes, some quite considerable, and for these, low sided wagons were used to limit the lift required in loading and unloading. One such wagon, illustrated here, was built by Gloucester Carriage and Wagon Co. in April 1898. It had brakes on one side only and No. 4 axleboxes, was 14 ft 11 in. x 7 ft 5 in. with 1 x 11 in. plank and was painted stone colour (buff) with black letters. Being registered by the GWR it probably normally ran between the Mendips or the Cotswolds and London and the South East.

Plates 62 & 63: Blast-furnace slag, the 'ashes', as it were, of the iron-making process, was a first class aggregate for road making, readily taking the binding agent due to its porosity. It was also stronger than limestone and did not polish easily. In addition it was convenient, for ironworks were to be found in areas which had no local supplies of roadstone or in the great conurbations. Here is not the place to discuss the varying chemical properties of different types of iron ore, which, of course, determined the properties of the slag, but suffice it to say that the slag produced on Teeside was admirable for road making. As a result, Major & Co., tar distillers of Hull, entered into an agreement with the Tees Bridge Iron Co. of Stockton, in 1913, and began making tarmacadam as the Middlesborough Slag Co. In 1919, Pease & Partners, the owners of the Tees Bridge Co., made a fresh agreement with Major's, who incorporated Tarslag Ltd the following year. Three years later this was floated as a public company, becoming Tarslag (1923) Ltd. Almost immediately old tips were acquired in the West Midlands, initially at Lightmoor, Corbyns Hill and Monmore Green, and in 1925 the headquarters of the company was moved to Wolverhampton. However, so large were the reserves that the Stockton tips were worked up to 1939.

Such was the widespread use of tarmacadam, that, at least in the 1920s and early 1930s, the railway wagon was the normal beast of burden, and it was customary to supply County Councils (the highway authorities) with a price per ton delivered to every railhead in their area. Being a prosperous and growing company, Tarslag (1923) Ltd invested heavily in new wagons, and one of the fleet, in this case built by the Derbyshire Carriage and Wagon Co. in 1927, is shown above. It is a 1923 RCH wagon fitted with side doors only and standard running gear, and is painted grey, with white letters and black shading. Lettered 'When empty return to Tees Bridge Works Stockton-on-Tees', these wagons carried a different livery to those built three years earlier by Charles Roberts, presumably for use based in the West Midlands, and shown opposite. The small lettering reads '*To be returned when empty to Tarslag (1923) Ltd., Slag Works, Monmore Green (GWR), Nr. Wolverhampton.*'

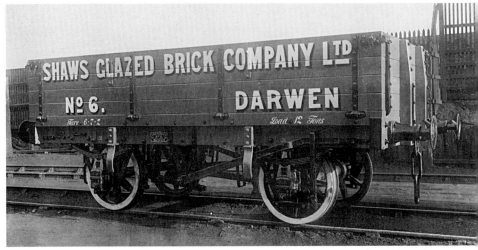

Plate 64: Before bricks were palletised and carried on special lorries with an integral crane, considerable tonnages were moved by rail. The bricks were stacked individually, with straw packing, and normally drop-side wagons were used to facilitate loading and unloading. As the load was heavy in relation to bulk, and liable to some movement in transit, such wagons were sturdily built. This picture shows a wagon from a batch of two, Nos. 5 & 6, built by Charles Roberts in 1926. It is fitted with twin door bang stops, normal for a drop-side wagon, but with the extra weight of a four-plank side, the stops are fixed in line with the face of the side rail, on metal brackets, and are reinforced with a washer plate at the top. On normal low-sided wagons the outer ends of the endsheeting are joined by a straight forward strap and a washer plate inside, but here a more substantial arrangement applies. An end knee, tapering from top to bottom, and bolted to the top of the solebar, ensures a solid corner to the wagon. The door fastening comprises a stud, similar to the solebar strap bolt, fitted to the top inside face of the end, which extends through a hole in the side. On the top plank is a plate, with a corresponding hole, which carries a conventional cotter and chain, the former dropping through a slot in the end of the stud. In addition the plate is extended round the corner of the wagon and into this engages a movable hook, pivoted to the top of the end knee. A real 'belt and braces' job. The wagon is 16 ft 6 in. x 8 ft 0 in., with 3 x 7 in. and 1 x 9 in. planks and appears to be painted grey (LMS shade), with white letters and black shading.

Chapter Seven

Liquids

Although numerically insignificant when compared with mineral wagons, tank wagons present by far the greatest variety and complex construction of all private owner wagon types, if not all railway wagons. Up to the 1880s liquids were usually carried in casks, earthenware jars or carboys. The rapid growth of the chemical industry, however, from the late 19th century onwards, led to a demand for the bulk carriage of liquids. Among the earliest calls for this was the transport of tar from gas works to chemical plants, for which the rectangular tank wagons were developed, and Charles Roberts continued to build these vehicles up to 1946. From the turn of the century, however, it was the cylindrical tank wagon which became the dominant type. The materials carried included petroleum and spirit, light oils, creosote, ammoniacal licquor, various acids, dye, various foodstuffs and more.

The design of tank wagons evolved over time, the main problem being the securing of the tank to the frame. The RCH produced various drawings to illustrate its recommended methods of fixing, but while these were broadly adopted they were not mandatory and any method could be used which was acceptable to the inspectorate. For many years timber was widely used, both for the underframe and tank mounting, by either the saddle or cradle method. In the former the tank was carried on four transverse bolsters (saddles), and in the latter it was supported on two longtitudinal members. Angle-iron end stanchions and wooden, or steel and wooden, horizontal beams held the tank longitudinally and the beams were diagonally stayed to the underframe by steel rods. The tank was finally secured by holding-down bands and crossed wire ropes. *Plate 67* shows a good example of a saddle mounted tank of this design. From about the middle 1920s the wire ropes were discontinued and a horizontal tie rod was introduced between the stanchion beams. Later on various bracket mounting techniques and welding were adopted, but these only apply to more modern vehicles beyond the scope of the present narrative.

Plate 65: (*opposite*) This view shows a similar type of wagon, built for Southwood, Jones & Co. Ltd, by Gloucester Carriage & Wagon Co. in January 1939. However, with only three planks to the side, conventional door stops and a straightforward pin and chain fastening were considered adequate. A more prominent difference, however, is that the outer limits of the ends are secured by fitting extra wooden stanchions. The wagon also has ¾ in. steel plates over the bottom. It is 16 ft 6 in. x 8 ft 0 in. with 3 x 7⅞ in. planks and is painted grey, with white letters and black shading. The small lettering reads 'EMPTY TO GRAIGDDU WORKS PONTNEWYNYDD GWR' (this was between Pontypool and Blaenavon).

Plate 66: In addition to tar, rectangular tank wagons were used for the conveyance of heavy oils. Of relatively simple rivetted construction, the tank sat directly on a wooden frame, between single, heavy duty end baulks, which were joined by tie-rods. It was held down by steel bands, with threaded ends, which located in brackets on the solebars. The example shown is one of two, numbered 3 & 4, built by Roberts in January 1913, and fitted with brakes on either side and oil 116 axleboxes. Two identical tanks, Nos. 1 & 2, were built two months earlier. It is painted black overall, with unshaded white letters.

Plate 67: (*previous page*) Tanks for 'Class A' highly inflammable liquids have been subject to strict controls since 1902. Regulations specified that the tanks must be cylindrical and that metal underframes must be used, but bottom outlets were allowed. However, after a number of near accidents involving leaking valves, an amendment was issued in September 1905, banning these and instructing that top vents and discharging pipes must be fitted.

In addition to their construction, the painting of such tanks was strictly controlled. The 1902 regulations specified that the tanks must be painted light stone (1930 British Standard 358, Light Buff is the exact shade), with a 6 in. wide Post Office Red (BS 538) band along the centre line. They also stated that the tank must carry the following inscription in bold characters, (here somewhat liberally applied in italics) '*No light to be brought near this tank. The cover and all other inlets and outlets must be kept securely fastened when not in use, whether the tank be full or empty.*' In March 1939, after pressure from owners, the RCH indicated that the tops of the frame, and the tanks, should be painted aluminium, with the red band on the end of the tank and a little way down each side, thus leaving more room for lettering and trademarks. However, this scheme was short-lived, for after trials with the Royal Air Force, it was withdrawn in 1941 and a livery less conspicuous from the air was introduced. This was matt dark lead grey for the tops of the frame and above. After the war the aluminium scheme was re-introduced, but the red band was dropped, although the inscription was retained.

The tank illustrated was built by Hurst, Nelson in 1916 and registered by the Caledonian Railway. When new it had oil 126-pattern axleboxes and the tank was 17 ft 5 in. x 5 ft 10 in. in diameter. The main name and end number was red, other lettering being black.

Plate 68: By the late 1920s the wire ropes had generally been discontinued, especially on tanks for inflammable liquids but the general principles of design remained as seen on this 14 ton tank built by Roberts in June 1930. It does vary from standard, however, as it is split into two equal compartments, which would allow different types of oil to be carried on the same journey. The register plate appears to read 'LMS', but other than this little is known about this company. the tank barrel is 17 ft 4½ in. x 6 ft 4¼ in. in diameter and the vehicle is painted black overall with unshaded white letters.

Plate 69: Being of a very specialised nature tank wagons were not taken over by the state in 1948, and new wagons continued to be built for private owners as this 1950 product from Roberts shows. Still based on the 1913 specification, the wagon has a capacity of 1,792 gallons, in a tank measuring 15 ft 4 in. x 5 ft 0¾ in. in diameter. On a standard 16 ft 6 in. frame this calls for the incurving end stanchions which give the vehicle its dumpy appearance. It is painted black overall, with white letters and the white star, 2 ft 0 in. across, denotes that it can run in fast goods trains (with an average speed of 35 mph).

Plate 70: One cannot help but equate the concept of a margarine wagon with Ken Dodd's Knotty Ash treacle mines, but the camera does not lie! Actually, of course, the tank was used for the carriage of the vegetable oil used in the manufacture of margarine, which was once the working man's alternative to butter. Being such an innocuous cargo the tank was no more than a rivetted tube with sealed ends, with a simple filler cap and discharge valve at the bottom. The provision of a ladder suggests that the tank was filled by a portable hose and pump, perhaps from a ship-borne tank. The pipework protruding from the lower end of the tank reveals that it was fitted with internal steam coils, to which low-pressure steam could be applied to heat the contents as an aid to offloading.

The wagon was one of a batch of 10 built by Roberts in 1925 and numbered 31–40. The presence of early pattern brake shoes and oil axleboxes indicates that second-hand underframes were used, although the Morton brakes were probably added when the tanks were built. The frame was 18 ft 0 in. over headstocks and 6 ft 10 in. wide, with a 10 ft 6 in. wheelbase and the tank was 17 ft 4 in. x 7 ft 2⅜ in. in diameter. It was black overall with unshaded white letters.

Chapter Eight

General Views

It is unlikely that any further substantial collections of official works' photographs will come to light, but there are views of private owner wagons still to be studied. Some of these are clearly seen in the general postcard views of stations, so beloved of many local photographers in the 1920s and 1930s. Others are contained in the archives and records of the former statutory bodies and nationalised industries and all manner of manufacturers. In other cases, which will be considered in the next chapter, part-views of wagons can often be all that will ever be known about certain vehicles.

Plate 71: It is an old adage that every picture tells a story and this view certainly does. At first glance it is a general view across the yard towards Thetford station, Norfolk, about 1920. However dominating the scene are three private owner wagons. The nearest belongs to the well known London coal factors, the Midland Coal & Cannel Co. It has side and end doors, the latter with pin and chain fastening, round based ribbed buffers, Ellis pattern axleboxes, and two wooden side door stops. It is 16 ft 0 in. x 7 ft 6 in. with 7 x 7 in. planks and is painted red, with white letters and black shading. It is numbered 505 and the small lettering below 'LONDON' reads *'Empty to New Monkton Collieries Nr. Barnsley.'* Beyond the sheeted wagon is a six-plank 10 ton wagon, with rounded ends, operated by Coote & Warren Ltd, Peterborough. It is fitted with side doors only, a single wooden door stop, ribbed buffers and united Ellis axleboxes. It carries the number 221, with a tare of 5-9 (on the bottom plank below the 'R' in 'PETERBORO') and is painted red, with white letters and black shading.

Finally is seen a five-plank, 10 ton wagon operated by Fosdick & Co., Ipswich. It is fitted with side and end doors and round bottom axleboxes. Measuring 16 ft 0 in. x 7 ft 6 in., with 5 x 9 in. planks, it is painted red, with white letters and black shading.

Ordinary enough wagons, but what are they doing in a country goods yard, 40 or more miles from their normal base? A major clue comes from the first wagon, lettered 'Empty to New Monkton', a colliery which produced only steam and gas coal. Reference to the Colliery Year Book and Coal Trades' Directory then reveals that one of the works of the British Gas Light Co. was at Thetford. There is little doubt, therefore, that the wagons illustrated are supplying coal to the gas works.

Plate 72: Charles Roberts built two batches of wagons for Fountain & Burnley, and although not photographed at Horbury, part of this company's fleet was caught on film at Barugh Green works, on the Silkstone branch, Barnsley, around 1930. Enlarged from a view probably taken to show the new building extension on the left, the shot shows (*left*) 10 ton wagon, No. 406. This has side, bottom and end doors, with pivot-bar fastening and Ellis axleboxes. It is 16 ft 0 in. x 7 ft 6 in., with 7 x 7 in. planks and is painted red, with white letters and black shading. Next is a 10 ton vehicle belonging to the well known coal factors and wagon builders, Hall Lewis & Co., Cardiff. It too has five doors, with raised end door roller bar and pivot fastening and united Ellis axleboxes. It is 16 ft 0 in. x 7 ft 6 in. with 4 x 9 in. and 1 x 7 in. planks and again is red, with white letters and black shading. Finally another 10 ton wagon, with five doors and raised roller bar. This wagon, however, has D-dropper end door fastening, internal diagonal side braces, hockey-stick bottoms to the side door and end knee washer plates and slender grease axleboxes similar to the MR type 8A. Again it is red and is 16 ft 0 in. x 7 ft 6 in., with 7 x 7 in. planks. For the record, and almost identical to wagon No. 406, except for having Attock's 46 axleboxes, those built by Roberts were: July 1904, 50 wagons Nos. 350-399, January 1905, 25 wagons, Nos. 325-349.

Plate 73: It is impossible to say whether this photograph was taken to show the battery of Becker coke ovens, or as a publicity photograph for E. & F. Beattie Ltd, Manchester, but it provides a valuable record of part of the latter's wagon fleet. The wagons are standard 1923 12 ton vehicles with side, end and bottom doors. The livery is particularly interesting, for examples of an advertising slogan taking dominance over the owner's name are very rare. The wagons are believed to have been grey, with white letters and black shading and the first two are numbered 15 and 20 respectively.

Plate 74: Despite the number of collieries in operation prior to World War II, photographs of them are rare. Add a few interesting wagons and they become very rare. Such is the case in this view of what British Coal Archives list as Coppice Colliery, taken in the early 1930s, probably to show the new loading screens. Confusion is caused, however, for while Coppice was operated by Shipley Collieries Ltd, the PO wagons belong to a neighbouring concern, the Manners Colliery Co. Ltd. The mystery deepens when it is revealed that Manners was no longer listed as a wagon owner by the RCH in October 1933, yet records show the colliery was worked up to 1949. One can only conclude that some sort of working or marketing co-operation was set up between the two companies.

Of the Manners' wagons visible, two are reasonably clear. In the centre of the screens is No. 623. This is a 10 ton wagon, with side and end doors (and probably bottom doors), with the end door bar above the end sheeting. It has plain, round based buffers but unfortunately the axleboxes are not visible. It is 16 ft 0 in. x 7 ft 6 in., with 6 x 7 in. and 1 x 9 in. planks. The white lettering is unshaded and this strongly suggests the wagons were painted black.

The second wagon under the screens, with diagonal lettering, No. 152, has side, end and bottom doors and square bottom grease axleboxes. It has ribbed round based buffers, but the most noticeable feature is the presence of two wooden door stops, and an early pattern metal one. It is 16 ft 0 in. x 7 ft 6 in., with 1 x 9 in., 3 x 7 in. and 1 x 9 in. planks.

The present narrative concerns itself with private owner wagons, but the two company wagons on the left cannot sensibly be overlooked. The nearest is a MR 8 ton wagon No. 126523, built to Diagram 299 in one of the lots of 1905 or before. In common with all D299 wagons it has side and bottom doors and Ellis patent axleboxes, and while it has not received its second set of brakes, it has been fitted with a later pattern door stop and striking plate. Photographs of D299 wagons in LMS livery are rare, this one being more so in that the letters 'LMS' were applied without a repaint, leaving remnants of the original 'MR' clearly visible.

Next to this is an equally interesting NER Diagram C10 High Goods, No. 26755, built by Hurst, Nelson in 1920. It has been modernised by the fitting of standard oil axleboxes and brake lever. Strangely it has been given twin door stops, with two striking plates on the door, replacing the normal, central door stop fitted to these wagons when built.

Chapter Nine

Part Views

If one compares the 1926 RCH list of private owners with the known photographs of PO wagons, it is clear that only a fraction of such vehicles are now recorded on film or glass plate. It is a sad fact that from a total of over 150 firms engaged in wagon building or repairing in the 1920s, official records of perhaps half a dozen remain. This means that any records, no matter how small, are very important. Such data come broadly in three forms, the part photograph, the contemporary sketch and the written record. Here we can consider the part view.

Plate 75: When Adams' North Staffordshire class 'New C' 0-6-4T No. 2042 was photographed at Stoke-on-Trent on 11th July, 1928, it was conveniently adjacent to Bassetts Ltd wagon No. 599. This was a 1923 RCH standard wagon fitted with side doors only. It was 16 ft 6 in. x 8 ft 0 in. with 5 x 7 in. planks, and is believed to have been painted black, with plain white letters, The letters 'LTD' follow the main name and 'ETRURIA' is on the right-hand end of the bottom plank. Being involved in the supply of tarmacadam these wagons would have travelled widely around the Potteries, Cheshire, south Lancashire and the West Midlands.

Plate 76: Lurking in the depths of Queen's Road goods depot, Sheffield, on 9th September, 1922, was this 5-plank 10 ton wagon operated by Wm Caudle & Co. This must have been a sizeable company and its owner believed in sharing his wagon building programme around the industry. His numbering policy also hints at visions of grandeur. Known wagons in the fleet are:

April 1893   Seven-plank 10 ton No. 20, 15 ft x 7 ft 6 in. x 4 ft 0 in. 2 ft 10 in. (5-plank) side doors. Built Midland Wagon Co.

November 1895   Five wagons, 15 ft 0 in. x 7 ft 6 in. x 3 ft 6 in., 6 x 7 in. planks, end rising 9 in., side doors only, No. 50 axleboxes. Painted MR Grey, with white letters and black shading. Nos. 60, 65, 70, 75, 80, built C. Roberts. Main name on 3rd-5th planks.

January 1897   As No. 20 but 3 ft 7 in. high and steel frame; Nos. 105, 110, 115, 120, 125.

November 1897   Five wagons, 15 ft 0 in. x 7 ft 6 in. x 3 ft 4 in., 2 ft 4 in. side door 4 x 7 in., 1 x 5 in., 1 x 7 in. planks; Nos. 85, 90, 95, 100, 130. Built E. Eastwood.

December 1899   The vehicle illustrated.
Five wagons, 15 ft 0 in. x 7 ft 6 in. x 3 ft, 2 ft 4 in. side door, 4 x 7 in., 1 x 9 in. planks, Nos. 27-31 (No. 27 broken up 1937, No. 31 - 1933). Built Harrison & Camm.

December 1907   One wagon, No. 7 as order of November 1897.

As a footnote this illustration stands as a testimony to the old glass plates, as this wagon was only 17 mm long on the original negative.

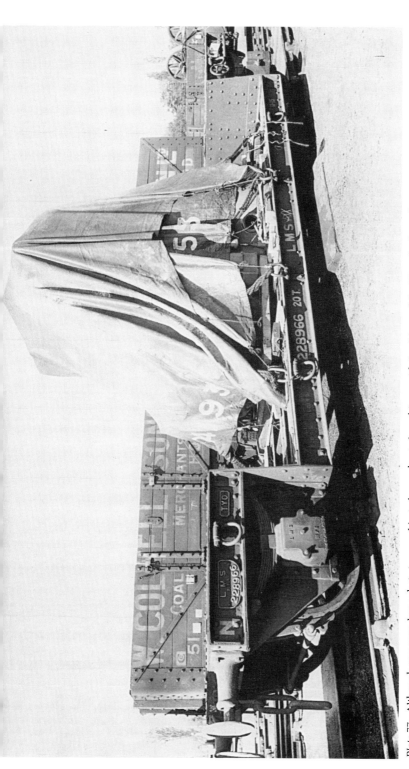

Plate 77: Although one wonders what strange object was sheeted on this LMS trolley wagon, seen at Northampton Castle, on 7th June, 1939, the real beauty of the photograph is the recording of the only known example of one of Wm Colwell's wagons from Hove. This is a five-plank, 10 ton wagon, with side doors only and Ellis patent axleboxes. It is 16 ft 0 in. x 7 ft 6 in., with 2 x 8 in. and 3 x 7 in. planks and appears to be painted red, with white letters and black shading. Wagons from the South Coast to the north were often routed via Clapham and West London line to Willesden. From Northampton it may well have gone on to Toton or Colwick via Market Harborough.

Plate 78: Most of the wagon photographs at Charles Roberts were taken in the yard to the east of the factory, adjacent to the Wakefield-Barnsley line, and quite a few passing wagons were captured for posterity. Such was the case in May 1938 when the LMS creosote tank No. 748900, was recorded (extreme bottom left corner). Rather more prominent in the enlargement is a 10 ton Glass Houghton wagon, from an unknown builder. It has side doors only (which probably means it was purchased second-hand), brakes on one side only and would have had grease axleboxes. It appears to be 15 ft 0 in. x 7 ft 6 in., with 3 x 9 in. and 2 x 11 in. planks, and is painted grey, with white letters and black shading. The wagon has a tare weight of 5-17 and on the side rail below 'CASTLEFORD' is 'NER & LYRS'. There are no apparent load details.

Plate 79: Seen between a locomotive smokebox door and a barely legible Wm Cory wagon (No. 4969) is a 10 ton wagon operated by John Mellor & Sons, Guide Bridge. This has side doors with cupboard style top doors and internal diagonal side braces. It would have had grease axleboxes but looks to be well maintained, which in itself suggests that its owner has complied with the requirements to fit the second set of brakes (Board of Trade Regulations, 1911). It is 16 ft 0 in. x 7 ft 6 in., with 7 x 7 in. planks and appears to be painted red, with white letters and black shading.

Chapter Ten

Contemporary Sketches

One of the most fortunate acquisitions of the author was the personal notebook of an unknown wagon repairer. This contains a large number of freehand drawings with dimensioned livery details. Made when a wagon came in for repair, in order that a full repaint could be given if necessary, these drawings do not give all details of the ironwork, nor any underframe particulars. On the other hand the majority of them show wagons which to the author's knowledge were never photographed. In a few cases it has been possible to relate the drawings to the wagon registers held at the Public Record Office, Kew or the NRM, and even to Charles Roberts' order books, and it is considered that the number of planks shown on each drawing is correct. One of the drawings is dated 1925, and the balance of the subjects suggests that the gentleman in question may have been employed at Wagon Repairs, Chesterfield.

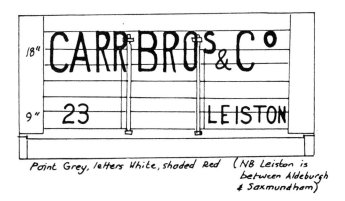

Point Grey, letters White, shaded Red (NB Leiston is between Aldeburgh & Saxmundham)

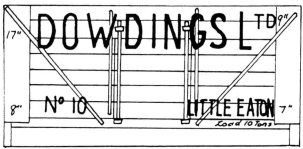

Painted Grey, White letters, Black shading

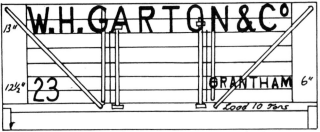

Painted Red Oxide, White letters, Black shading

Loaded to Thomas's Sidings   For Repairs advise
        GWR Bourne End   Wagon Repairs Ltd, Gloucester

Painted Dark Red, letters White, shaded black
    Paint date 2/27   Cost of lettering 7/3½ᵈ (38p)
                              Painted at Hirwaun, Glam.

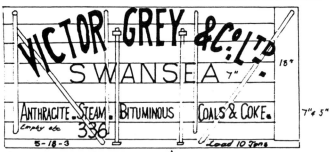

Empty to Raven Colliery   Painted Red, letters white
    Pantyffynon, G.W.R.   shaded black date 3/27
        South Wales   Cost of lettering 6/11½ᵈ (35p)
                                   Painted at Hirwaun, Glam.

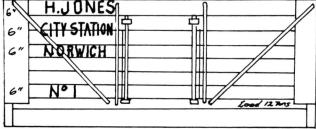

Painted Slate Grey, White letters, Black shading
    All letters 6"

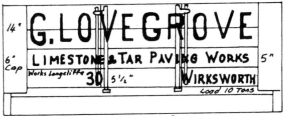

Painted Red, White letters, Black shading

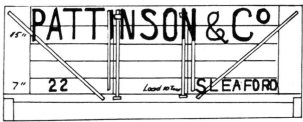

Painted grey, White letters, Black shading

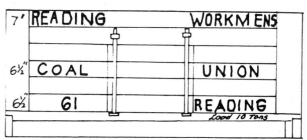

Paint Black, White letters, unshaded

Painted Red, White letters, Black shading

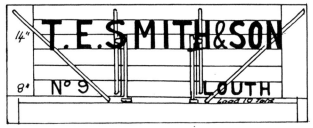

Point Red, Letter White, shade Black

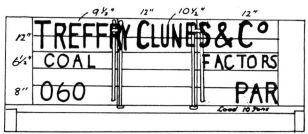

Point Red, Letters White, Shaded Black

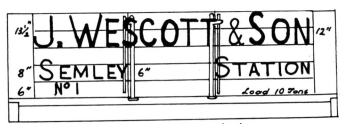

Point Grey, White letters, Black shading
NB Semley is between Salisbury & Templecombe

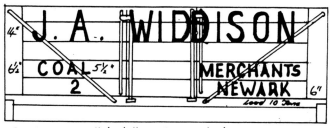

Pointed Red, White letters, Black shading

Chapter Eleven

The Written Record

Probably the most important documents found by the author resulted from that once in a lifetime occurrence of being 'in the right place at the right time'. Purely by accident, research led me to Charles Roberts, Wakefield, just at the time the National Railway Museum was being developed, and eventually I became a voluntary 'rescue body' on behalf of the museum. One day, while sorting through a mountain of drawings, I was approached and greeted with, 'Av' got something as might interest thee'. Introducing himself as a foreman, I was taken to part of the erecting shop and under a desk I was shown a large hessian sack. Inside were a number of huge bound ledgers, which were the works' order books from 13th April, 1893 to 17th April, 1923. 'I found em in an old cupboard last year, are they worth owt?' After assuring him he was the most wonderful man I had ever met I set out for York with the priceless sack. It took me 12 months to get there, because that was how long it took to copy the orders, by hand, into a chronological sequence.

When one peruses these orders it becomes clear that many wagons left the works without facing the camera, but the written details should provide sufficient detail for all but the most fastidious modeller. The following selection has been made, again covering wagons of which the author has never seen a photograph. Dimensions given are internal, sheeting is the thickness of the planks. All ironwork normally painted black.

21st November, 1893   J. Banks, Hillhouse (Huddersfield)
2 wagons 14 ft 6 in. x 7 ft 0 in. x 3 ft 6 in., 6 x 7 in. planks, 2½ in. sheeting, 2 side doors
No. 50 axleboxes. Red, white letters, black shading. Nos. 1 & 2

JOSEPH BANKS
COAL
B
MERCHANT
No1
HILLHOUSE

30th, May 1894   Shillito & Burnley, Huddersfield
2 wagons 14 ft 6 in. x 7 ft 0 in. x 3 ft 6 in., 6 x 7 in. planks, 2½ in. sheeting, 2S, 2B doors
No. 50 axleboxes. Red, top three planks on sides white. Main name black, red shading.
Other lettering white, shaded black Nos.1 & 2

SHILLITO & BURNLEY
COAL
HUDDERSFIELD
MERCHANT
No1

July 1896   T.S. Hanson, Blyton (Nr. Gainsborough)
1 wagon 14 ft 6 in. x 7 ft 0 in. x 2 ft 11 in., 5 x 7 in. planks, 2½ in. sheeting, 2S to top, 2B
Ellis axleboxes. Red, white letters, black shading. No. 1.

T.S. HANSON
COAL   MERCHANT
No1   BLYTON

14th August, 1897   J. Moore, Askham-in-Furness
1 wagon 14 ft 6 in. x 7 ft 0 in. x 3 ft 6 in., 6 x 7 in. planks, 2S
Ellis axleboxes. Dark brown, white letters, black shading. No.1.
Deliver new to Nostell Colliery.

JOHN MOORE
COAL   MERCHANT
No1   ASKHAM-IN-FURNESS

9th February, 1900   T.H. Harvey, Plymouth
1 rectangular tank 15 ft 0 in. x 7 ft 1 in. x 3 ft 4 in.
Attock's 58 axleboxes. Black white letters. No.7.

T.H. HARVEY
CHEMICAL WORKS
No7   PLYMOUTH

12th September, 1900   One wagon as above. No. 8.

23rd October, 1902   Cadwallader, Liverpool
2 wagons 14 ft 6 in. x 7 ft 0 in. x 4 ft 0 in., 7 x 7 in. planks, 3 in. sheeting, 2S, 1E
Attock's 46a axleboxes. Ruby red, white letters, black shading. Nos.1&2 (on door).

CADWALLADER   20"
CROWN ST   I   LIVERPOOL

24th August, 1909   Haagensen Watt & Co., Hull
5 wagons 15 ft 6 in. x 7 ft 0 in. x 4 ft 0 in., 3 x 9 in., 1 x 8 in., 2 x 7 in. planks, 3 in. sheeting,
2S, 2B, 1E
Attock's 46a axleboxes. Removable coke grating 1 ft 10 in. deep, side door springs.
Tarred inside. Green, white letters, black shading.

3rd May, 1910   C.H. Appleyard, Mirfield
1 wagon 15 ft 6 in. x 7 ft 0 in. x 4 ft 0 in., 7 x 7 in. planks, 3 in. sheeting, 2S
Attock's 46a axleboxes. side door springs. Red, white letters, black shading. No. 1.

C.H. APPLEYARD & SON
MINERAL WATER MANUFACTURERS
No1   MIRFIELD

31st August, 1911   W. Brook & Son
2 x 12 ton wagons 16 ft 0 in. x 7 ft 5 in. x 4 ft 3 in., 5 x 7 in., 1 x 8 in., 1 x 9 in. planks, 3 in.
sheeting, 2S, 2B.
Oil 116 axleboxes. Side door thresholds. Nos. 4&5.
Red, white letters, black shading, blue band, white edge & letters.

SLAITHWAITE
4   AND   HONLEY
WILLIAM BROOK & SONS   DYEWORKS

20th September, 1912   J. Metcalf, Accrington
1 wagons 16 ft 0 in. x 7 ft 0 in. x 2 ft 3 in., 3 x 9 in., 3 in sheeting, 2S to top
Oil 116 axleboxes. Side door springs, brakes each side.
Dark red, yellow letters, black shading. No. 11.

JOHN METCALF
ALTHAM
No11   ACCRINGTON

22nd January, 1915   Wm Fowler, Norwich
10 x 12 ton wagons 16 ft 0 in. x 7 ft 5 in. x 4 ft 3 in., 5 x 7 in., 1 x 8 in., 1 x 9 in. planks, 3 in. sheeting, 2S, 2B, 1E.
Attock's 158a axleboxes. Side door springs, brakes each side.
Red, white letters, black shading. Nos. 301-310.

W. FOWLER

No   NORWICH

Small letters each side 'If empty without labels please forward to Pinxton Colliery, via Midland Railway'

Chapter Twelve

How long did they last?

This is a question often posed by modellers of the early BR scene, and is one which is not easy to answer with great accuracy. It is known, for example, that some of the ICI lime wagons built by Gloucester Railway Carriage and Wagon Co. in 1937 were still in main line service up to about 1962/63. Just prior to this a large number of Clay Cross Company wagons could be seen in the sidings around the station of that name. What is known is that in 1958 the British Transport Commission announced that the railway's Modernisation Plan called for a substantial reduction in the size of the wagon fleet. In pursuance of this policy a list was compiled covering 160,000 wagons which were to be considered as surplus to requirements. This list was over and above those already condemned and included 6,220 former private owner wagons which had not been renumbered with the 'P' prefix. Many of these wagons were shunted away in collieries, yards and sidings and while in excess of 6,000 seems a lot, not many appear to have been photographed. The remaining three illustrations give a glimpse of what these wagons looked like, and give a challenge to the *aficionados* of the weathering technique. Good modelling!

Plate 80: This standard 1923 wagon, with a magnificent load of pea slack, which has actually been flattened as the wagon left the screens, was captured at Grimethorpe Colliery about 1948/49. It was originally painted grey, with white letters and black shading, but only the latter colour on the ironwork and shading remains to any meaningful extent. The white has virtually gone, except in the left hand diagonal, but traces of the body colour remain where the lettering stood. The wagon itself looks in sound condition and was probably taken into internal user by the NCB, where it could have lasted into the 1970s.

Plate 81: Seen at Cudworth on 24th August, 1952, this 12 ton wagon built by W.H. Davis about 1922/23, was virtually in original condition. It has side, end and bottom doors, and Attock's 58a axleboxes. It retains the original 'handed' brake shoe at the right. The original red body colour has almost gone and the new, unpainted planks on the door, have weathered to a dirty brown.

Plate 82: Seen shortly after withdrawal at Clay Cross on 10th June, 1967, this 15 ton wagon was built by Thomas Moy in 1913 and fitted with side, end and bottom doors and Ellis patent axleboxes. When new it was red, with unshaded white letters, with the word 'MAKERS' on the second door plank, and 'FUEL ECONOMISERS' on the top plank. It was not originally numbered 1106, which was a six-plank wagon built by Harrison & Camm in 1909, but was from a batch numbered 1181-1280 built to carry iron ore.